Living with a Stranger

John Stewart Collis

Living with a Stranger

A Discourse on the Human Body

'Men go abroad to wonder at the height of
mountains, at the huge waves of the sea, at
the long courses of the rivers, at the vast
compass of the ocean, at the circular motion
of the stars; and they pass by themselves
without wondering.'

Saint Augustine

GEORGE BRAZILLER • NEW YORK

Published 1979 by George Braziller, Inc.
Originally published 1978 by Macdonald and Jane's Publishers Ltd, London
Copyright © 1978 John Stewart Collis

Library of Congress Cataloging in Publication Data
Collis, John Stewart, 1900–
Living with a stranger.
Bibliography: p. 191
Includes index.
1. Body, Human. 2. Human physiology. I. Title.
QP38.C64   612   78–21307
ISBN 0–8076–0912–9

First edition.
Printed in the United States of America.

To my wife

Contents

✳

Prologue

❦

DURING MANY YEARS now I have deplored my ignorance concerning the body. I would hear people say: 'My liver is sluggish,' or 'I have got kidney trouble,' or 'His blood pressure is too high,' or 'She has a weak heart,' or 'You are as old as your arteries,' or 'I have caught a bug,' and I would feel ashamed of how little I knew about these matters. I am a person who knows what he doesn't know, and I have been truly disconcerted to hear someone tell me that his liver is out of order, while not knowing on my part what the liver actually is; or, for that matter, what blood is composed of and why we need it and how it circulates in a 'stream'; or how it happens that we carry about in our bodies a temperature about equal to that of the tropics, regardless of whether it is hot or cold outside. I live with this body of mine, and yet for all I know about it I might be living with a stranger.

It may happen, if I write this book for myself in order to clear up what I want to know and do not understand, that there will be others who, acknowledging a similar ignorance and a like desire to overcome it, may wish to join me. If so, they may feel as I do that the subject has been overburdened with nomenclature even more than botany itself. An excessive 'naming of parts' is not entirely beneficial and can be injurious to comprehension. I will name parts. I may name many parts: but only those which I wish to remember and am concerned to pass on. The reader may also share with me a distaste for physiological illustrations: with Fig. 1, Fig. 2, Fig. 3, and arrows pointing to various parts of muscular nudes or skeletons. I do not think I was ever made sad by Burton's

Anatomy of Melancholy; but when turning over the leaves of Gray's famous book I have been so filled with the melancholy of anatomy that I have not had the heart to pursue my studies. It should be enough on this occasion to visualize without 'visual aid' a human body, place the figure before us and see if we can realize what it is.

It seemed to me that my best plan would be to approach the subject in the same way, from the same angle, and with the same technique as I applied to my subjects in *The Vision of Glory* and *The Worm Forgives the Plough*. In order to be sure of my facts in those books, if I found that I could not follow the authorities, I would button-hole friends whose speciality in those subjects made them fair game. When they saw me approaching they would say to themselves: 'Here comes Collis about to ask one of his *simple* questions,' and they would often take to their heels. Not so with my friends Dr Stephen Kane and Dr Judy Kane when I approached them with medical questions. They stood firm. They did not flinch. I would ask a question couched in terms of crude metaphor. They would say: 'It is not quite as simple as that'; but they did not thereupon make it more complex, they gradually cleared it up for me; and I hope it will not impede their careers if I openly thank them and say that I doubt whether I would have attempted this enterprise without their help, for while they were indulgent to what they called 'Collisian audacities' they did their best to save me from gross error.

We Nourish Ourselves

❧

I   We breathe

LET US THEN REGARD the figure of a man standing in a field (since we must think of him as stationed somewhere). We go close and look at him carefully. The first thing we notice about him is that he is *breathing*.

Why is he doing this, we might well ask ourselves. We do not ask it, because to breathe is so much part of our daily experience. We know that while we can do without water for several days, and without food for many weeks, we will die in a few minutes if we fail to breathe. What is so terrible about a corpse is not the pallor in the face or the stillness of the limbs, but the *not-breathing*.

So we are bound to ask – what then is breath? And we must answer: it is food. It is the essential element in our nourishment. (Once, when attending a vegetarian luncheon party, I sat next to a woman who refused every course – explaining that she was on a diet of air.) And when we ask why we should take nourishment of any sort, why we need what we call food, we face the fact that no living thing is a separate entity, but is part of other things and therefore must partake of them in order to exist. To some extent a stone is a separate entity, receiving nothing, and giving nothing away save that which may be exacted by the erosion of time. But see that cow in the field, with its big wet nose. It is part of the field, and must continually be recruited from the field – which we call eating; but if it fails to do so it will sink down into the field – which we call dying. In his *Elementary Physiology* T. H. Huxley makes a careful experiment after which he is entitled to claim that in the course of an hour a living active man 'undergoes a *loss of*

11

substance'. He adds: 'Plainly, this state of things could not continue for an unlimited period, or the man would dwindle to nothing. But long before this diminution of substance become apparent to a bystander, it is felt by the subject of the experiment in the form of two imperious sensations called hunger and thirst.'

But more imperious is the desire for air, since our bodies are chiefly composed of gas made concrete. It is natural and proper that we should call air air, and not gas, especially that we should refer to 'fresh air', for such is our experience; but since it is a composition we are bound to take note of the fact that it is composed of nitrogen, oxygen and carbon. We do not see these ingredients in the air; they are not apparent, they are transparent – until suddenly conditions alter and the combination of elements gives us vestures of cloud or showers of rain. The oxygen element is the main thing when we breathe. This is what we must have if we are to live. It is the essence which is laid as the foundation of all the living world, and without which the earth would lie ghastly and bare like the plains and mountains of the moon. Everything that grows must have oxygen in the first place, and if it is to continue living, whether moving on the surface of the land, flying above it, or swimming in the depths of the sea, must always be supplied with fresh draughts of it. What then *is* oxygen? We do not know. We have clapped a name upon an essence in order to conceal our ignorance. Best to leave it there.

How did this man whom we have pictured standing in the field come into existence? Who are his ancestors? Jane Carlyle once wrote to her friend Mrs Russell: 'If there is one thing I dislike more than theology it is geology. When Darwin in a book that all the scientific world is in an ecstasy over, proved the other day that we are all come from shell-fish, it didn't move me to the slightest curiosity whether we are or not. I did not feel that the slightest light would be thrown on my practical life for me, by having it ever so logically made out that my first ancestor, millions of millions of ages back, had been, or even had not been, an oyster.' I would go along with that, even substituting the words 'philosophic life' for 'practical life'. I admit to a measure of curiosity as to whether our ladder of ascent was from the creatures of the forest or the creatures of the sea – or both, as seems more likely. But I think it is important to recognize the unimportance of these questions, and to go much further back to find the father – in fact to the amoeba. No, further still – to the boiling soup before the forming of the earthly crust.

'Men of Science like to think,' wrote Macneile Dixon in his great work, *The Human Situation,* 'that the genes are physico-chemical structures, and that life arose in a chemical ferment in the first, or protogene, through the action upon it of cosmic rays, an interesting hypothesis, not less probable, indeed, than that it was the work of an angel. In the absence of evidence all speculations are of approximately equal value, and you can take your choice.' In fact our choice is valueless, since it is bound to be fanciful. It is best to stick to what we know and come to the main concern. We know that when the earth's crust was formed there was inorganic matter. We understand today that while inorganic matter is not alive it is very lively. This stone which I hold in my hand is composed of millions of boxes in which charges of electric energy are spinning, a whirlwind of electrons revolving in their orbits millions of times in the millionth of a second. This liveliness within the stone is not at all obvious to us; if we put it to our ears we hear nothing. It is therefore very difficult to realize that all substances are composed of atoms which are, actually, empty rooms enclosing the dance of the electrons and the nucleus which is really an atom within the atom. That was the first stage in the story. I am not able to explain, for I do not understand, how the first germ emerged from these chemical sources. 'To develop to our level from the chemical soup,' writes John Taylor in *Black Holes,* 'is thought to have taken several billion years.' I am very ready to accept this and to accept that it was a continuity of evolvement, since that is more conceivable than the idea that the first germ appeared without ancestry.

When Bishop Wilberforce, in a famous debate, enquired of his opponent, T. H. Huxley, 'whether it was through his grandfather or grandmother that he claimed descent from a monkey?' Huxley whispered to a friend sitting beside him: *'He has delivered himself into my hands.'* For Huxley, in his riposte to this *bon mot,* after a careful review regarding the evidence for evolution, concluded by declaring that he 'would rather have an ape for his ancestor than an intellectual prostitute like the Bishop.' I would go further. I am free to say that I would prefer to have a germ or a microbe or a bacterium as my ancestor than that kind of bishop. In any case I have no choice. If we are to consider ancestry at all we are obliged to go back to the germ. I often hear people speaking unkindly of these things. 'I have caught a bug' a man will say, assuming that I will commiserate with him for having been the victim of something bad. But I cannot help thinking that bugs and bacteria and microbes and the like* must intrinsically play a beneficial part in the economy and balance of nature, and can only do us harm if we ourselves have in some way uneased the balance and made ourselves vulnerable to disease: and I take it that Bernard Shaw had some such consideration in mind when in *Too True to be Good* he represented the Microbe as being pained by the attitude of the patient, claiming that it was she who had made *him* ill.

Anyway, I think we are obliged at least to claim the amoeba-like creature as our ancestor. One thing is clear. As soon as what we may call the Living Principle was established then the everyday forces of the inorganic world became its servants. The dance of the electrons subserved the cells. The entirely physical energy, in bottomless and unboundaried hoard, could

---

*Are viruses living things? If so, must we regard them as being outside the primal order, simply devils prowling up and down seeking whom they may devour?

14

be channelled by the Principle of Life into all the forms we know of growth and motion. When the Principle departs the process is reversed and the inorganic reigns supreme. Then the electrons control the empty envelope. Then, as T. H. Huxley said, 'oxygen, the slave of the living organism, becomes the lord of the dead body.' The discarded corpse, abandoned by the constructive spirit, must surrender to disintegration. No longer able to control the laws of mechanics and chemistry and consume the environment in plant and animal, it must dissolve. No longer armed to battle against bacteria and confront the elemental erosions, it must waste away. Though wasted it is not lost. Its atoms return to earth and sky to appear sooner or later as the bricks laid to the foundation of other organisms, so that the atoms that informed a monkey in the Stone Age may later become part of the brain of Julius Caesar, while 'Imperious Caesar, dead and turned to clay/Might stop a hole to keep the wind away.'

II Cells of life

THE AMOEBA is generally considered as the first living thing superimposed upon the whirling electronic activity of the atoms. The electrons are not alive in the sense that they can eat each other or multiply themselves. The amoeba introduces the new Idea: matter which is composed not only of electricity but also of gas made visible in a given conditioned shape and capable of receiving more gas which likewise turns into concrete substance. We are referring to the first germ-cell. Just as a bit of matter is not a solid (as you might conceive a lump of jelly) but is composed of millions of atoms, so also a living bit of matter, a body, is composed of millions of what we call cells. To be alive, to be a creature, as opposed to a piece of matter, it is sufficient to have only *one* cell. This is the amoeba. A cell is a sort of box with a number of remarkable things inside it. We cannot make atoms visible, but by means of the microscope we can see what a cell is like. We can immensely magnify it and throw its picture upon a screen. Having done so we find a whole world in itself with globules, filaments, vesicles, all in perpetual agitation as each fulfils a particular function – a whirlpool of snake-like gliding and darting movements in interlocked association. And in the midst a nucleus, and in the nucleus chromosomes, and within the chromosomes the genes. These genes – another word which we use to cloak our ignorance of final mystery – are responsible for what the creature becomes, and are handed down from generation to generation so that we speak of genealogy.

Handed down? Yes, for given one cell, then two cells or two million became possible because the machinery within the cell promotes the absorption of matter outside it (at first chiefly water and oxygen) which in turn promotes growth and the power to *divide* and again and again divide till from the single-celled amoeba to man the number of cells rises to billions. It is this capacity to absorb, to grow, and to multiply – for here division is another word for multiplication – that enthrones the primal cell which emerged from the primeval soup as monarch of the living world. It is claimed that if nothing had hindered the sub-divisions of the amoeba, its descendants would in a week have equalled the earth in size. Of this creature, amoeba, Jennings made the just observation: 'If it were the size of a dog, instead of being microscopic, no one would deny to its actions the name of intelligence.'

The first thing that the amoeba has to do in terms of absorption is to take in oxygen. This is quite easy. All that is necessary is contact between its surface and the oxygen of the air. The cell simply lies in the air, or in water containing air, and allows the oxygen to do its work. I do not propose at this point to write three hundred pages concerning the biological evolution of the species, but will return to the finished article we started with – the man standing in the field whose first necessity was to breathe. It is not sufficient for him just to let the air beat against his body, because instead of one cell easily permeable to the oxygen, it has to reach the millions of cells that compose every inch of his person. The brain alone accommodates a frightening number of cells and if even for a few minutes they get too little oxygen it is a serious matter – we have all experienced on such occasions the horrible feeling of dizziness.

Thus the man standing there is obliged to take in a considerable amount of air every minute.

Do we not therefore have, we might fairly ask, an immense network of air tubes or channels for distribution throughout the body? No, there is no necessity for air-pipes *per se*. Since the body requires a great deal more nourishment than oxygen, it can be delivered to us in pipes which are full of liquid. So we come to *blood*.

III   The digestive system

THE CELLS of our bodies are like so many million mouths hungrily feasting upon ingredients carried to them in the streams of blood.

Since the ingredients are prepared in the alimentary canal we must first consider its cargo and its work. Indeed, we should start with the mouth, but as we are fully aware of what our mouths are and do, there is no need for description. But it would be pleasant to pause, if only for a moment, at the chief property there, the teeth, white glittering rocks growing in the soft flesh, fashioned in the calcium foundry and built up into enamelled fangs and grindstones. In the animal kingdom they often do more than the office of grinding; they serve as aids in locomotion, as a means of anchorage, as instruments for the uprooting of trees, as transport for building materials, and most emphatically as weapons for offence and defence. Beholding the bared teeth of a tiger the celebrated huntsman, Selous, declared himself bewitched. It is not the crocodile's tears but the long rows of terrible teeth in the sly face that frighten us. For humanity their appeal is also aesthetic, saving a plain face or ruining a beautiful one, while at all times our manner of speech is enhanced or degraded by their properties. It is not surprising that in some countries they have been worshipped as idols and treated with religious awe. The Kandy Tooth was understood to be a relic of the Buddha, and was the most famous of Buddhist antiquities. For over a thousand years in Ceylon it was one of the most valuable vehicles of occult virtue, to be compared with the Cross, the Nails and the Blood. Indeed, the Tooth was the most potent idol of the Indies, and if a king could lay claim to it and style himself Master of the Tooth he would be regarded as the greatest king in the world and supreme arbiter of the Eightfold Path; and on one occasion when the Tooth had been stolen, a sum of £1,000,000 was paid for its restoration.

The task of our teeth is to break up our food sufficiently well that we may swallow it with ease. But what exactly is this stuff which we call food and are supposed to take in order to keep on living?

We are reminded of Henri Fabre's story about the French chef who having prepared an elaborate meal was offended by a man who declared that he knew someone who could produce it all quite easily out of three bottles, one containing air, one with nothing but water, the third with a bit of carbon. The aggrieved cook, on enquiring who this conjurer might be, was told that 'that supreme artist, the green vegetable cell, given a portion of hydrogen, nitrogen, oxygen, and carbon, can, through means of sunlight, build a bacillus, a tree, a mouse, or a man.'

We are part of the earth and part of the sun, and so in order to keep going we must continuously partake of the sun and of the earth. Since the general composition of the body remains constant it follows that our food must be identical with our bodies. We are composed of a few minerals and gases – elements which can be chemically compounded into so many different forms. So long as we can command a constant supply of these minerals and these gases we have nothing to worry about.

We need substances which fall under four heads: proteins, fats, carbohydrates, and minerals.

Proteins are composed of four elements: carbon, hydrogen, oxygen, and nitrogen, sometimes united with sulphur and phosphorus. These are made manifest as the chief constituent of muscle and flesh.

Fats are composed of carbon, hydrogen, and oxygen only – but with an excess of hydrogen. These are made manifest as fatty tissue and oils.

Carbohydrates also consist of carbon, hydrogen, and oxygen only – but with much less hydrogen. These are made manifest in starch and sugar.

The minerals comprise sulphur, phosphorus, potassium, sodium, calcium, magnesium, silicon, iron, and chlorine.

Now since, like that French cook, we are such experts in the use of these basic materials to produce a variety of tasty food, we appear to be in a very superior position in this matter of supplying ourselves with nourishment. Yet in comparison with the animals we are at a disadvantage. Our man standing in the field is handicapped in relation to the cow or horse beside him. True, they are also dependent upon the activity of that vegetable cell and its operation of photosynthesis without which life would not only halt but never start. But that horse and that cow can partake of the field directly while we must eat the cow in order to eat the grass: we are all vegetarians of course, but many of us take our veg. in the form of meat.

However, we find no real difficulty in laying hold of these essential ingredients of which our bodies are composed, though even so we are still at a disadvantage in comparison with animals, since before putting the food into our mouths we are generally obliged to soften it. Nothing need be said here about the art of cooking. I admit to surprise at some of our methods: thus, we know that if we heat a lump of fat we thereby turn it into a liquid which then helps to soften substances. But when we boil an egg its liquid contents become hard. We eat it that way, transforming it back into fluid state, almost as it were uncooking the cooked and de-solidifying the solid. We might with advantage have taken it 'raw' in the first place, even if only as a measure in economy. I mention this just in passing and by way of coming to our next consideration which is the work of the alimentary canal in liquidizing solids and distributing the gases and minerals which we need.

The first movement occurs in the mouth when the teeth undertake the preliminary task of transforming material from one thing into another thing – and I put it that way deliberately because we shall see that the principle in the canal is continuously to turn one substance into another substance. The teeth begin the work by chewing the material. We recognize, because we feel, the importance of this action, and the dental industry rests upon our appreciation of it, for though teeth seem as hard as marble they are as frail as flowers, in swift subjection to corruption and erosion. We are assured on high authority that Gladstone did *not* chew every morsel thirty-two times, just as Marie Antoinette did not ask, 'Then why don't they eat cake?', but the ideal that gave rise to the Gladstonian legend remains laudable. The teeth, in combination with a fluid in the mouth, saliva, break up and liquefy the morsels, after which by means of that slippery mobile platform, the tongue, which can be raised to the roof of the mouth, we are enabled to swallow them. The food descends, not by gravitation, but by a muscular action in the throat so powerful that we could eat or drink while standing on our heads. This is carried out very consciously, but the moment we have swallowed the stuff the work is out of our control and we are obliged to leave it to the engines of our interior until the rejected portions reach the rectum for expulsion, when we become acutely aware of the success or failure of the expedition. That such operations should continue with so much success into or beyond a person's ninetieth year argues notable labour worthy of our attention.

When we swallow the food we pass it into a tube known as the alimentary tract or canal composed of a substance called gut (so far-reaching in its virtue that the word has overtones in music, in tennis, and in courage). This tube goes from the throat to the bowels, some 30 to 40 feet in all. The stomach forms a part of this canal in the same way as we might see in the mind's eye a small lake as part of a river. It is not necessarily the most important part, for at a pinch we could manage without a stomach, but it is best to possess one because it is a muscular bag in which the churning of food carries further the preliminary work of the teeth, and passes it on in a still more friable state to the intestines. It is also useful as a larder. In the evolution of species it has further served as a kind of cloakroom, for when food is 'bolted' it is advisable to keep it somewhere in order to ease the breaking-down process. This seems especially true of the aquatic creatures up to the present day, some of whose stomachs are apparently hospitable to indigestible matter, and I have recently heard of a shark found near a caye off Kingston, Jamaica, in whose stomach were discovered a camera, a bunch of keys, a chest-of-drawers, and a pair of trousers.

We must always speak respectfully of the stomach. Professor Alex Hill, in his immensely learned and detailed volume, *The Body at Work,* permits himself an occasional literary caper (thereby convincing us all the more of the reliability of his knowledge). 'My lord the stomach! He is not the only, nor is he the chief, agent in digestion; but with him rests the decision as to whether the food offered to the alimentary tract is suitable in quality and quantity. He is offended if it be not offered with all the circumstance and ceremony which becomes his rank. As an intimation that he is about to receive food, he accepts the news from the mouth that its nerve-endings are subject to mechanical stimulation. But the chewing of india-rubber would produce a like effect. The stomach therefore confers with the organs of taste and smell. If their report is favourable, he argues that the substance which the teeth are crushing will justify an outflow of gastric juice. He responds most generously when prolonged mastication assures him that he may trust to receiving the food in a sufficiently subdivided state. At our peril we neglect to propitiate my lord. Not always debonair when treated with consideration, he is morose or petulant when slighted. Never content with lip-service, he exacts the labour of teeth and tongue and palate. The tribute we offer may be of the best – savoury, wholesome, well cooked, well chewed – but if it be not tendered with some degree of love, if thoughts are concentrated on other things, if no attention is devoted to the meal, if no sense of liking accompanies our offering, my lord the stomach on his part affords the viands an indifferent reception.'

I second that. For perhaps the worst offender in failure to propitiate my lord is your literary man who too frequently bolts his food, and is guilty of a tendency to be thinking of something else. Had I been aware of the intricate work being carried out on my behalf down there, I would have led a better life and been a better man.

Note the phrase 'justify the flow of gastric juice'. The first step in the digestive operations is the promotion of saliva called forth by taste and smell, to ease the work of the teeth. Saliva is a remarkable fluid, alkaline and watery, rich in sodium, potassium, chloride, and bicarbonate; cessation of salivary secretion can lead to foul breath within an hour and dental decay within a week. Then, in turn, the descent of the food through the throat (more properly, the oesophagus) calls forth gastric juice in the stomach. This is a very powerful lotion, hydrochloric acid, so virulent in its action that it could burn a hole in the carpet. Indeed, the question has been raised – why does it not eat away the gut itself? I understand that it would do so, and actually does do so, if not otherwise fully engaged. This juice is supplied by a gland (we might define a gland as a kind of flesh purse) that contains this fluid which remains dormant until the presence of food calls it into action. Once in operation it acts as a further solvent and breaking-down agent working in combination with three layers of muscles in the stomach which churn the materials. This gastric juice contains a chemical property called pepsin, which is another word for an enzyme, which is a general term applied to a substance which causes a change in another substance. The most important office of the gastric juice is its enzymatic (or catalytic) effect upon protein.

Thus the food having first been chewed by the teeth, churned by the stomach, and treated by the gastric acids, has already been altered so much that we now give it another name – chyme. It is beginning to assume the appearance of pea soup.

Yet its journey is far from over, nor its treatment completed. After a few hours in the abdominal bag it passes on – as chyme – to the next station which stands at the entrance of the small intestine, a resting place called the duodenum. And now it is exposed to another juice, another attack even more drastic than the assault of the hydrochloric acid. This issues from another gland called the pancreas. It is extremely versatile in its capacity to turn one thing into another thing, for it also is an enzyme. It turns starch into maltose, and maltose into glucose. It splits fat and curdles milk. It furthers the preliminary work of the gastric juice by turning proteins into peptones, the peptones into polypeptides, and the polypeptides into amino acids – smaller fragments each time.

It is important to forget these names while acknowledging the existence of an organic laboratory where a combination of grinding and liquidizing turns part of a mutton chop, or a potato, or a plate of salad, into absorbable matter leaving the rest to be rejected and passed on. Thus the work in the alimentary tract is a straightforward process: the saliva in the mouth promoting digestion in the stomach, and the juice in the stomach stimulating the secretion of the pancreas.

It seems that of all the stations through which the food travels the small intestine provides the most action, whereas the large intestine is more in the nature of a drainage system than a laboratory. In giving this account of the digestive system I have not sought to stretch my description on the rack of a too-detailed completeness; but I am persuaded that it is comprehensive enough to save me from the impropriety of an excessive brevity. And we are now free to consider how the finished article, the chyme, is absorbed into the blood stream.

IV The composition of blood

THE TERM blood stream cannot easily be envisaged in the singular. It is difficult enough to envisage the Gulf Stream, a river of water running through banks of water from Mexico to Scotland. One would expect so much diffusion that it would soon fail to qualify as a stream. Still, that strange phenomenon does in some such sense exist. But not a blood stream in the singular. What we have *not* come to in this story is a prepared fluid of nourishing essence reaching a given harbour, as it were, whence it flows into a channel called the blood stream.

Our bodies are absolutely suffused with blood. It is vital to us. We can do without this or that organ, but not without blood, our 'life-blood' as we justly say; for if it pours from a wound unchecked, life will pour away with it. 'From the wound his life's blood was spouting in a stream, and falling with a hiss into the road,' says Thomas Hardy describing the death of Tess's horse, Prince. 'In her despair Tess sprang forward and put her hand upon the hole,' a gesture as natural as it was naïve, and she was splashed with blood as she stood helplessly looking on. 'Prince also stood firm and motionless as long as he could; till he suddenly sank down in a heap.' The amount of blood on the road surprised Tess: a quantity which had also surprised Lady Macbeth in another context: 'Who would have thought the old man to have so much blood in him?' she jabbered in her sleep.

Luckily the composition of blood is such that a wound seldom leads to much loss, and a cut is of little account. For blood is not homogeneous like water. In a drop of water there is only water. In a drop of blood there are particles termed corpuscles which are of two kinds; one lot are red, the other white or colourless. The red are far more numerous than the white, and the name haemoglobin is used to denote this, a remarkable compound having the largest molecule of any known organic substance. It contains iron, and it is this that gives the blood its colour. 'Is it not strange,' asked Ruskin, 'to find this stern and strong metal mingled so delicately in our human life that we cannot even blush without its help?'

If a given quantity of blood is poured into a basin, it is at first perfectly fluid; but in a few minutes becomes congealed into a jelly-like mass, so compact that the basin could be turned upside down without any of the blood being spilt . At first it is a uniform red jelly, but soon drops of a clear yellowish watery-looking fluid make their appearance on the surface. These drops increase in number and run together until it becomes apparent that the jelly has separated into two different constituents: the one a clear yellowish liquid; the other a red semi-solid mass which lies in the liquid. The first is called the *serum*; the semi-solid mass is the clot, and it contains the corpuscles of the blood bound together with fibrous-looking matter which we call the *fibrin*. When blood flows from a cut it is the fibrin which forms the plug, nature's plaster, that prevents further flow, and we call it coagulation – more swift and effective in hot weather than in cold.

This coagulation is a great thing, as we all know from experience, since an ordinary cut is of little consequence, and a bad one will coagulate with the aid of a bandage. (It is obviously best to allow a light cut to congeal without assistance, and to wait until the clot has formed before sponging away the blood that has trickled over the skin: the cut heals quicker that way, in my experience, for healing comes from without as well as from within.) In rare cases the blood does not congeal and a person impeded by this defect in his organization will be in danger throughout his life of bleeding to death after the slightest wound. If Tsar Nicholas's son had not suffered from haemophilia, the historians of the Russian Revolution would not be obliged to include the deplorable episode of Rasputin.

It is this coagulation which has had such an effect upon the consciousness of mankind with regard to blood. For it causes a *stain* which does not easily wash away; water cannot dislodge it, nor scent dispel its odour – '*all the perfumes of Arabia will not sweeten this little hand.*' Yet sometimes the lasting power of a blood stain is a thing to cherish. The tunic which Nelson wore at the Battle of Trafalgar is becoming brittle; the cloth is wearing out, but the guardians of the coat declare that the blood stain caused by his fatal wound, lovingly preserved, is the strongest portion of the garment, and will outlast the fabric.

When a terrible wound is sustained the copious haemorrhage which follows often induces fainting. This is a good thing, for as the organism slows down, the clot has time to form – so if a man faints at the sight of his own blood, it is really the best thing that he can do. Women have less tendency towards this reflex, partly through natural fortitude and menstrual experience, and because for centuries in the past it was the rôle of women to dress wounds rather than receive them.

The facts about blood are always arresting. Apart from its watery portion, called plasma, which is just over 50 per cent of any given amount, there are the corpuscles. The plasma contains as its base a combination of carbon, oxygen, nitrogen, and hydrogen (the four elements which preponderate in our structures), together with the minerals which are also necessary to us. Yet within each pinprick of blood there are also the red corpuscles (or cells), white corpuscles, and platelets. The red ones number five million, the white ten thousand, and the platelets two hundred and fifty thousand – all in the one pinprick. It has been calculated that the blood cells in a human body, if arranged 'shoulder to shoulder' in single file, would stretch over two hundred thousand miles –

more than two-thirds of the way to the moon.

These cells are alive, as much alive as the amoeba, and are themselves composed of atoms and molecules: there are 280 million molecules of haemoglobin in each red corpuscle, every molecule containing ten thousand atoms. And since they are alive they are subject to death, their life-span being between three weeks and three months; but as they constantly die they are as constantly replaced.

The function of the red cells is to continually replenish oxygen for the body, as I will soon spell out in detail. The function of the white cells is to attack bacteria and microbes which may be proving harmful to the organism. I write 'may be' for we cannot suppose that the body is full of microbes and bacteria perpetually at war with us, as doctors and physiologists tend to suggest. They must serve in other ways more often. Anyway, when they do assault us and make us feel ill, the white cells assault *them* and eat them up. In this they generally succeed. Thus when we are sick and feel terrible we should take comfort from the thought that we have an interior army fighting for us, which understands the engagement and will win the campaign. These white cells are sometimes called leucocytes and sometimes phagocytes. The eating up of the germs which are against the body – not to be confused with 'antibodies' – is known as phagocytosis, and Lord Lister once made the memorable observation that 'If ever there has been a romantic chapter in the history of pathology, it is the story of phagocytosis.'

Of course the leucocytes or phagocytes can and do die in their thousands in battle with bacterial toxins, and the pus in the stye of an eye, for example, is a manifestation of their death; but, speaking for myself, I am always glad to see the pus, for it generally means that though the phagocytes are dead their victory is assured, and when we have squeezed out the pus (or graveyard) we are nearly back to normal. In the event of our own interior resources being insufficient to overcome the enemy we call in medical aid. In *The Doctor's*

Dilemma, Shaw represents one of the doctors, 'B.B.', as having one cure-all, which was 'stimulating the phagocytes'. But since the complexity of the organism is so great, Shaw depicted 'B.B.' as having a bee in his bonnet, and instead reserved approval for the doctor called Ridgeon, based on Sir Almroth Wright, who had discovered certain substances in the blood which he named *opsonins* that in some way affect microbes and render them more palatable and attractive to the phagocytes. Ridgeon knew how to do it: 'B.B.' was only a man with a slogan.

How about the third lot of cells – the blood platelets? They are smaller than the red cells and the white, but more numerous than the latter. They are really broken-off fragments of the larger cells, but they play a very positive function in the assistance of coagulation, fortifying the fibrin in the clotting of blood at the site of the injury, and thus achieve an honoured place in the harmony of the metabolism.

The average person contains about nine to twelve pints of blood, say a pint for every stone of the body. This does not seem excessive, yet for many centuries and in most civilizations until quite recently it was thought too much. *Bleeding* had been considered as a sovereign cure for all sorts of things: nervous excitement, fever, malevolence, malaise. It was sufficient for a man to have a red face to qualify him for a prescription of bleeding by fashionable medicos, while many a poor wretch suffering from nothing worse than feeble fibrin would be mercilessly bled by brutal quacks. It is only fair to say that the practice was also favoured by monarchs impressed by the opinion of Hippocrates and the arguments of medieval sages, so that we find Frederick the Great demanding that his veins be opened during battles in order to soothe his nerves, while Louis XIII of France accepted forty-seven bleedings in six months, and Charles II of England submitted to useless bleedings even on his death bed. The practice died hard, for even when it became evident that the cutting of blood-vessels encouraged septicaemia a new idea was introduced, as gruesome as that of the legendary vampires – the employment of leeches. A profitable trade was made in leeches hired out to fasten on your skin and suck your flesh, costing half-a-crown for half an ounce per leech. This practice was especially popular in Paris at the beginning of the last century when a great leech enthusiast, Dr Broussais, was alleged to have been instrumental in the shedding of twenty million pints of blood in France alone.*

The tide has now turned. Any loss of blood is regarded as potentially serious, and should be replaced. The idea now is to pour it in, not out. For this purpose blood-banks called transfusion centres are organized and donors deposit a pint of their blood, duly registered according to its 'group'. Thus the wheel turns in medicine as in so many things. Yet 'wheel' is inaccurate; medicine does not go in circles but advances by short rushes. There are not only blood-banks for those with too little, but serum-banks to be drawn upon by those whose blood cells have not the strength to win the day against a given attack, but can be reinforced by the injection of a specific serum. But I must not look through this window and endanger the brevity of my brief.

V   The circulation of blood

THE SUFFUSION of blood throughout the body is contained in a conglomeration of tubes we call veins, arteries, and capillaries, the last being so numerous that it would seem our tissues are enmeshed in a network of tubes more closely knit than a spider's web. So closely knit, in fact, that there is not a single place on our bodies, except our nails, hair, and teeth, which we can prick with a needle without pricking a capillary and drawing blood. The tissues of the body might fairly be described as the fabric that holds the capillaries together.

---

*See *The Body* by Anthony Smith.

This is where my difficulty comes in over talking about the blood *stream*. We appear to be confronted with an inextricable network of interlocking tubes, each containing a stream. If we magnify any bit of our bodies, we would see an apparently endless labyrinth of waterways. 'A lot of holes surrounded by stuff' as a boy once described a net. A condensation of elm-tree tracery. But there is no labyrinth. There is no confusion. Nothing is entangled. These blood-vessels always carry the liquid in one direction only, and working in concert, flow one into another, all taking a circular route. The design is perfect (physiology reeks with purpose and design). And since our streams of blood do all flow in one direction, we are at liberty to accept the general term of blood stream.

For hundreds of years, indeed until the seventeenth century, very little was known about what the blood was or did. Because it was not known what the heart was or did. Aristotle thought that the heart was a central heating furnace and that the arteries were filled with air that cooled it. There were all sorts of notions. Was the blood stationary? Did it somehow ebb and flow? How was it channelled? It was not until 1615 that Dr William Harvey came forward, and concentrating his attention, not upon the blood but upon the heart, cleared up the matter. For twelve years he carried out dissections, not only upon man and the larger animals but also upon frogs and lizards and doves, on oysters and tortoises and snails, on crabs and slugs and shrimps, on mussels and serpents and fishes, even on wasps and hornets and flies: after which he felt entitled to claim that 'it is absolutely necessary to conclude that the blood in the animal body is impelled in a circle and is in a state of ceaseless motion, that this is the act and function which the heart performs by means of its pulse, and that is the sole and only end of the motion and contraction of the heart.' Harvey had discovered that the blood *circulated*. Now every schoolboy knows about the 'circulation of the blood' without ever saying to a schoolmaster, 'Isn't it pretty clever of the blood to do that, Sir?' Yet it is certainly surprising. It was too much for

Harvey's contemporaries. After the publication of his book the doctor lost prestige. 'He fell mightily in his practice, 'twas believed by the vulgar that he was crackbrained.' There was an extraordinarily conservative desire to cling to the opinions of Galen, a second-century physiologist who knew nothing regarding circulation, while a prominent professor of medicine in Paris, Guy Patin, declared that Harvey's theory was as useless as it was absurd.

However, Harvey did receive recognition during his lifetime though it was not until five years after his death that the microscope came to confirm it decisively. The great Dutch microscopist, Leeuwenhoek, who saw capillaries in the tails of tadpoles, wrote: 'I saw that not only the blood in many places was conveyed through exceedingly minute vessels from the middle of the tail towards the edges, but that each of the vessels had a curve, or turning, and carried the blood back towards the tail, in order to be conveyed back to the heart.' There were many who thought that Leeuwenhoek was only a conjuror performing a trick with glass, but it was soon generally accepted that the invisible could be made visible in this way, and the existence of capillaries must be acknowledged. But the *amount* of them was hard to accept, since it was claimed that if those in one human body were laid out in a straight line they would stretch far enough to cross the Atlantic Ocean. And how could the blood circulate neatly in such a network?

Since then the problem has been explained in many a learned treatise. A short, clear, general statement may be attempted here. We are dealing with two organs – the heart and the lungs – which work together. And really with two blood circulations: one with reference to the lungs, the other with reference to the rest of the body. The heart, situated to the left of the chest, is a pear-shaped organ about the size of a man's fist, and is a bag containing four compartments. The lungs surround it: in a bold figure we could say (and I have seen illustrations that prompt it) that the heart and lungs, viewed together, resemble a spaniel's head: the heart being the face and the lungs being the ears. The covering of the heart is all muscle, with the impulse to continually contract. The compartments are full of blood which is in a perpetual state of motion because of that contraction of the heart muscle which ejects blood in one direction and instantly receives a similar amount to fill the vacuum. This contraction is generally likened to a pump and is popularly called a beat. From sixty to seventy of these compulsive beats (which can be felt pulsing in an artery) occur every minute of our lives. This works out as two thousand million beats in an average life-span.

The blood from the body first enters the heart by an ear-shaped entrance called the right auricle and is then pumped into another chamber which, looking like a little stomach, is called the right ventricle. This ventricle, using a second pump, a pump of its own as it were, forces the blood around the lungs from which it receives abundant oxygen and flows back into the left auricle, and from there into the left ventricle. At this point the pumping system is more powerful than hitherto for now it ejects all the heart's blood into a main artery called the aorta which conducts the blood into all the arteries, big and small, and through the little arteries into the capillaries, and then through the capillaries into the little and big veins which finally end in two big veins (the *venae cavae*) that open into the right auricle where we started. This circular journey of the blood takes about fifty seconds.

How is it that the blood never gets entangled in the network of tubes, or runs backwards? It does not become entangled because it is all propelled the same way towards the larger vessels, the veins, which lead directly back to the heart. It does not run backwards because it is not like a river which can only flow downwards. It can flow upwards equally well, and any back movement is checked by a series of valves which, while favouring its forward flow, impede a backward one. And we should note that arteries are much more muscular than veins, and contract more actively, thus forcing the blood in the one direction.

The work of the heart begins when it is microscopically

small. Harvey described how he saw the new-born heart beating in the embryo of a chicken. He added: 'I have also observed the first rudiments of the chick in the course of the fourth or fifth day of incubation, in the guise of a little cloud, the shell having been removed and the egg immersed in clear tepid water. In the midst of the cloudlet in question there was a little bloody point, so small that it disappeared during the contraction and escaped the sight, but in the relaxation it reappeared again, red and like the point of a pin; so that betwixt the visible and the invisible, betwixt being and not being, as it were, it gave by its pulses a kind of representation of the commencement of life.' Yes, a symbol indeed! A mere beadlet in the midst of matter; a pin-point in a cloud; yet its beat spelling the assurance of life as its failure the sentence of death. It was always recognized as such long before the actual circulation of blood was pronounced. 'May I shake your hand,' said a young man in the street to the aged Cervantes, 'you, Sire, the pride and joy of the Muses.' Cervantes replied: 'You have met me at an uncouth hour. My life is slipping away, for by the diary my pulse is keeping, which at its latest will end its reckoning next Sunday, I must close my life's account.'

The full-grown heart is a tough organ, as indeed it must be, and to have 'a weak heart' is too loose a phrase to be accepted with solemnity. It can be handled without damage and even changed for another. It can be taken out of the body of an animal such as a frog, or of a mammal, and if given nutrition by being immersed in a solution of salt and sugar can be kept beating for days, as if it were a whole live animal rather than the organ of a dead one.

The heart may be tough but it is highly susceptible to the mental state of its owner. Emotion affects its motion. When Charles Dickens was touring the country to give his famous recitations from his novels, his pulse-rate alarmed the doctors in attendance, for its beat went to danger point, and when he played the death of Nancy at the hands of Bill Sikes his pulse rate rose from seventy to one hundred and twenty. And we all know how fear shows itself on the face. A murderer bending over the body of his victim looks up as the door opens – to reveal a policeman. Immediately his countenance becomes 'as white as a sheet', for his heart has almost stopped beating and the red corpuscles have been drained from the capillaries of his face. On the other hand if we make some foolish mistake and are very embarrassed our heart will beat faster and blood will flow more than normally to the cheeks, and we will blush.

The heart needs to be tough and powerful because every time the left ventricle contracts it drives into the aorta about four tablespoons of blood at a velocity of about a hundred feet per minute. William Paley claimed that the whale's aorta was larger than the main pipe of the waterworks then at London Bridge, and that the blood rushed through it with greater impetus than through that pipe. Now, if blood is propelled through the body at such a rate, what good is it, we might ask, in terms of distributing nourishment to the tissues? The answer is that it is only propelled along at that rate at the beginning. It soon slows down – for we must allow for the vast subdivision of the innumerable vessels through which it subsequently passes after leaving the aorta. 'If the water pipes supplying a town branched until the original circuit was represented by five to six thousand million little pipes,' writes Dr Alex Hill in a striking image, 'the friction which the pumping station would have to overcome would be very great. But little force would remain in the water when it reached the smallest pipe. Still greater is the resistance to the flow of blood, which is slightly viscous and contains solid corpuscles which increase friction. Two thousand miles of capillary tubing in the body of a man, without reckoning his liver and lungs!'

VI   The vascular and lymphatic systems

WE HAVE NOW seen what the blood is composed of, how it circulates, how it acquires oxygen from the lungs and how its flow is regulated by the operations of the heart. We have not yet seen how it receives nourishment from the digestive system and how the chyme makes its entrance into the veins, arteries, and capillaries – that is, into what we call the vascular system. Is there a special place in the alimentary canal from where it is distributed, a kind of port, a Plymouth, where the goods are loaded into the aorta? It is not like that. Thousands of capillaries join the canal all the way down and as the food passes along, digestion and absorption go hand in hand; but it is especially in the small intestine that the chyme is taken up by the blood vessels and passed into the whole arterial network. In a broad way we might call the small intestine the chief port of departure.

Yet still we must ask, if the veins, arteries, and capillaries are tubes enclosing blood, how can the tissues of the body receive any of the contents of these tubes? The answer is that they receive none from the veins and little from the larger arteries, but there is no problem in receiving it from the capillaries. The circumstance that all the tissues are outside the vessels by no means interferes with their being bathed by the fluid which is inside the vessels. In fact the walls of the capillaries are so exceedingly thin that their fluid contents readily exude through the delicate membrane of which they are composed, and irrigate the tissues in which they lie. Thus the picture we are entitled to draw is a propulsion of blood sent in waves throughout the body at roughly seventy times every minute, spreading through all the branches of the vascular system, and by means of the multitudinous capillaries nourishing the tissues according to their respective requirements.

Yet this vascular system requires support from *another* set of capillaries, another elaborate system of canals. For besides the capillary network which constitutes the vascular scheme, all parts of the body need lymphatic channels which act in concert with the main design.

What is the purpose of this additional network, what part does it play in the economy of the whole?

The fact is that the blood capillaries are unable to prevent the serum from seeping out too much and thus giving an excess of fluid to the tissues. It is the office of the lymphatic system to take up this excess of serum and put it back into circulation. Since its vessels are highly permeable it is able to absorb this fluid and eventually pour its contents into the venous system at several sites, the most important of which is via the thoracic duct which empties into the left jugular vein at the side of the neck. Thus gross congestion in the tissues is avoided.

This is not the only service performed by the lymphatic system. The interstitial fluid picked up by the lymphatic vessels is not all good stuff. It contains a certain amount of waste matter, foreign material, dead cells, particles of dirt, bacteria, and so on. This noxious material which could cause disease is disposed of and 'treated' by a complex system of filters called lymph nodes which are also responsible for the production of lymphocytes that are part of the body's circulating defence system.

Thus the lymphatics not only return the surplus serum to the circulation but also purify it. Moreover they supply about 40 per cent of the white blood cells in the body, which form part of the freight ceaselessly arriving at the thoracic duct. The system can be thought of in terms of a factory, a vehicle and a scavenger.

In summary, we have two networks. First we have the vascular scheme carrying out the circulation of the blood – a very elaborate design as we have seen. Intermingled with this we have the secondary network of canals which we call the lymphatic system. And when we think of the lymphatic system we must think of four things: the lymphatics, the lymph, the lymphocytes, and the lymph nodes. The lymphatics are the channels; the lymph is the fluid in those channels; the lymphocytes are the white cells in that fluid; and the lymph nodes are filters and factory centres producing those lymphocytes which are discharged into the lymphatics and finally disgorged into the thoracic duct.

We should note, in passing, that more than nourishment is involved when man uses the vascular system to help himself in other ways. Though subject to far more diseases than the animals, he can cope with them to an extraordinary extent by benefit of his medical skill. I get a pain in my toe or my ear, or both. A doctor gives me a pill to swallow. (I am speaking of the expert medical man.) It contains ingredients which if mingled with my blood will ease that pain. He sends the aid by vascular canal to the afflicted spot, with good effect. It will not reach and act upon the injury in fifty seconds, for it must first be diffused through the whole network (though he can send it by express, as it were, with a direct injection). The pain-killer pill is only a small example regarding our capacity to employ the vascular system for sending messengers of mercy and comfort. What is so fascinating for the layman is the diagnostical expertness by virtue of which modern doctors can use the circulating blood to *cure* a given disease and not merely alleviate pain, by sending through the blood stream a formidable variety of substances which according to necessity act beneficially upon the injury. This is why the doctor is entitled to stand outside the general run of mankind, and, like the artist, perform apparent miracles. No man is held in greater esteem than the doctor. All power is based upon *need*: he is powerful because we need his help. It has always struck me as an interesting gloss on religion that the medical acts of Jesus have commanded such attention, though healing, *per se*, has no theological significance.

VII   Lungs, warmth, energy

THE READER will have grasped that there are really two blood circulations, and that the pulmonary one which deals with the lungs is vitally important since oxygen is so basic to existence. There is no difficulty in the blood abstracting a sufficient quantity of oxygen from the air in its rapid passage through the lungs because every day we receive into and expel from our lungs four thousand gallons of air, and the surface of the air-vessels totals approximately one thousand square feet. The lungs are essentially a collection of very small bladders around which the blood flows, each air-cell being a cup of thin membrane holding together a basket-work of capillary vessels of which there are said to be seven hundred million in the lungs, their total surface in single file enough to stretch from London to New York. This vast surface in the aggregate presents no problem to the red corpuscles in the blood since they themselves are so numerous that in single file, we are told, they could stretch for two hundred thousand miles. Such calculations contain an element of the absurd: forget them, while taking the point that there is a lot of air coming in, a lot of channels to receive it, and plenty of corpuscles to benefit by it. We are acutely aware of the necessity of oxygen, since fresh air is so prized by us and foul air so hateful – and lack of air lethal as in the famous Black Hole of Calcutta, a room in which one hundred and forty-six people were locked with only two small windows, the consequent loss of oxygen causing the death of all save twenty-three.

There is an additional reason why we require this gas. We need it in order to burn. The temperature inside our bodies must be kept constant. We are 'hot-blooded' creatures, and even the 'cold-blooded' animal needs some heat. Our temperatures keep a steady 98° to 100° Fahrenheit whether we are in the tropics or at the North Pole. It is true that we can burn to death or be frozen to death, but this is the result of an attack from without and bears no relation to the fire within.

The creeping coldness as death extends its icy hand owes nothing to the outward state: 'So a'bade me lay more clothes on his feet,' said Mistress Quickly in her imperishable report. 'I put my hand into the bed and felt them, and they were as cold as any stone: then I felt to the knees, and they were as cold as any stone, and so upward and upward, and all was as cold as any stone.'

It is difficult to grasp what heat *is*, beyond giving a chemical explanation. The combination of carbon and oxygen generates heat, for some reason. If conditions are such that the molecules move very fast this seems to cause what we call heat. A wet haystack is burning. If it burns too much the hay will adopt another vesture in the form of flame, and the stack will be seen to be consuming itself, as it were, and will depart from the field in smoke. There was a large hayrick in a field on a farm where I once worked, which had been assembled while the grass was too damp. The farmer feared that we would be in danger of losing it so we climbed on top and dug a deep hole through its middle. I was interested to discover how oven-hot it was, and I jumped into the hole. Had my mates left me there and covered over the top (as they threatened to do) I could have been burnt alive. Whenever carbon is exposed to oxygen there is burning. To burn does not necessitate flame. The prostrate tree-trunk in the sad neglected wood sinks slowly into ashes as the burning bark decays. Fire is the result of *rapid* union of carbon with oxygen. The temperature of our bodies is produced by the effete carbon of the system united with the oxygen taken into the blood at the lungs. Thus we are composed of masses of little torches; in every crevice of our frame is a kind of languid fire, or slow combustion, ceaselessly at work, steadfastly burning from day to day and from year to year. Go into a cold empty hall and then wait until it is filled with three hundred people. It is not surprising that in a fairly short time the place is very much warmer – for three hundred fires are burning within it. 'I warmed both hands before the fire of life,' said Landor; 'it sinks, and I am ready to depart.' To be young is to contain more fire than later on. Strange essence! 'I frequently turned into my bunk soaking with wet,' wrote Herman Melville concerning his his first voyage when he was working his passage to England, 'and turned out again piping hot and smoking like a roasted sirloin.' How baffling also the imperatives of mind upon our inner heat! We become more hot if flustered or made foolish; we shiver when afraid; and if our spirits are high the outer temperature may pass unheeded: when Lenin mounted his sleigh to return from exile in Siberia, he declared that his inner warmth was so great that he did not need an overcoat.

Do we not stand in danger sometimes of swift combustion, ourselves turning into flame like a haystack? In *Bleak House*, Charles Dickens killed off the bad character called Krook by 'spontaneous combustion', and in the preface to the bound edition of the book answered those who had questioned the possibility of such a death with a listed report of similar incidents. Even today there are those who delight in claiming instances of such occurrences. Michael Harrison, in his *Fire From Heaven* (1976) tells us of Mrs Euphemia Johnson who returning home to Sydenham in 1922 to make herself a cup of tea was found that evening as a mass of calcined bones 'which were lying within her *unburned* clothes', while the chair on which she had been sitting was unscorched. Mr Harrison assures us that combustion need not be as fatal as in the case of Mrs Johnson, and he cites the case of a Lincolnshire servant-girl who erupted into 'wing flames' easily extinguished by a neighbouring farmer; and he refers to a Mr H. Nashville who was able to stifle a long gas-jet of flame which shot out of his trousered leg, without burning the cloth. If it is unnecessary to take such tales too seriously, we are assured on good authority that inflammatory gases as eructations have caused explosions at surgical operations, and I learn from a surgeon that gases from a cow's anus recently ignited a farm building in Shropshire, in consequence of which the vet was successfully sued by the farmer. I am happy to add that

Anthony Smith makes it clear that the true story of a
certain parson whose breath caught fire each time he blew
out the altar candles was due neither to drink nor spontaneous combustion but to gases proceeding from an ulcerous
stomach.

The properties in the blood and the operations of the heart
and lungs do not only bestow nourishment and warmth upon
us. They give us strength. They give us energy. 'I cannot fly
into the sun,' said Carlyle, 'but I can lift my arms – which is
equally strange.' It is indeed. It might not seem to follow that
just because we have had something to eat, because we have
taken into ourselves a portion of our environment, we should
thereby be enabled to *move*. Yet so it is. When Einstein
announced the equation $E = mc^2$, he was pointing out that
energy and mass are forms of the same thing: that mass is
energy made manifest, that energy is invisible mass. When we
are young we hungrily suck up nourishment to build our
bodies and we rush about using the energy supplied with the
material. When we grow up we are inclined to eat more
(having greater opportunity) and rush about less, and
sometimes become almost stationary. In this case we are in
danger of adding to our mass the mass which should have been
used as energy. The result is often unnecessary or unseemly
bulk. The protruding stomach of the fat man is, as it were,
bottled energy preserved as mass. It was easy for him to
acquire this. It is much less easy for him to turn his mass
back into energy and lose bulk. But of course there is no
question that he can do so if he wishes. He need only starve
himself for a period. We might ask why, if he simply eats
nothing, he should grow thin. A dead body, if shielded from
bacteria and not eaten by vultures, does not waste away for a
long time and some have been preserved for thousands of
years. We waste away and become terribly and even gro-
tesquely thin if we truly are starved, because then the body
feeds upon itself, the most dedicated vegetarian becomes a
flesh-eater, even a cannibal, though the person he eats is

himself. He can continue to excrete, but that which he is depositing is part of himself. 'It has been rightly enough observed,' wrote T. H. Huxley, 'that a starving sheep is as much a carnivore as a lion.'

VIII   Excretion and kidneys

IF WE BEND LEVEL with the grass dew-drenched in early morning, we see the endless spider-webs spread over all the phalanxed spears. We love to see them. We never tire of the streams, the rivers and the rivulets of the earth. But the bit of earth we call ourselves is a stranger to us. Nothing could be more wonderful than the vast design of hair-sized tubes we call capillaries, but our acknowledgement is only in disgust when a tube is pricked and blood is seen – blood, the living liquid that we hate to see, and use only in the language of a curse. I think of this just now because I have not yet finished with this aspect of our story. We have been concerned only with what has been received, and not with what has been cast away. So I must come to *faeces*.

We have already remarked that in order to exist on the earth we must take part of the earth into ourselves; and even when we die and go under the earth (having been looked after by the undertaker) we give ourselves back to it. That is fairly obvious I suppose. What is not so obvious is why, having nourished ourselves as a precaution against wasting away, we should discard a good deal of that food, and, indeed, call it a 'waste-product'. The fact is we are unable to turn everything completely into chyme and are obliged to discard some of it. Part of our mutton chop or salad dish must be rejected. There is no problem about doing so: we simply eject it.

But there does seem to me here a rather striking difference between ourselves and the animal world in general. I mean in terms of purity, innocence, wholesomeness. Take the elephant. In its case (perhaps in any case at all) we have no right to speak of 'waste matter'. The material taken in by the elephant is shifted out in large quantities as nutritious in another mode. Dr Malcolm Coe tells how in the Tsavo game park a young elephant will produce about 8 lbs of dung every time it defecates, while a large one would produce up to 60 lbs which would be repeated fifteen to sixteen times a day – that is, some 900 lbs. Then the dung beetles get to work on it, and mixing it with the earth greatly increase the fertility of the soil. 'The movement of soil by termites,' adds Dr Malcolm Coe, 'is an aid to fertility since like earth worms in temperate regions they aerate the soil and incorporate food material. The same is true of the dung beetles which turn the soil over and in so doing not only aerate but also incorporate the dung so that when it decomposes below ground the nutrients are available to plants. When you think of up to thirty thousand elephants defecating fifteen times a day, the amount of soil shifted is astounding.' Thus the elephant excretes much of what it has eaten; but that which has been rejected by it is accepted by plants, which in their turn grow up to be eaten by the animal whose excretion they ate.

Take the polyp. Its exudation is responsible for the formation of coral. By virtue of their multitudinous excretions the coral insects build up reefs whose rocks often resemble brilliant flowers, and can comprise an area of 33,000 square miles, or create coral islands of extravagant beauty. Take the sea-birds who deposit guano. Long before Europeans realized its value to agriculture because of its extraordinary richness in phosphate the Incas of Peru forbade anyone under pain of death to visit the guano islands during the breeding season. Or take the familiar farmyard manure which is so highly prized by us all. We recognize the purity of its soil-nourishing effects. I remember as a small boy being impressed, while in a pony trap, by the way the horse during the drive would sometimes excrete with such completeness and competence that absolutely nothing was left at the clean pink anus.

How different with us! Let the most beautiful woman or

man sit down at a table and eat a perfectly cooked meal with delightful odours proceeding from it. Subsequently that man and that woman will deposit a given amount of the food in what is felt to be so disgusting a state that we make it in secret in a small room designed to get it out of sight as quickly as possible. Even so it is still not 'waste material' in the economy of nature. If we happen to make a deposit, say on a summer morning in a wood, in the clear clean air, it will only be a matter of seconds before it is entirely covered with flies which lap it up as if it were strawberries and cream. It is certainly not waste material for them; and we are bound to acknowledge that flies play their part in the general scheme of nature as certainly as bacteria – though not necessarily with an eye to human advantage.

The teasing question remains as to why human excretions are so much less pure (and useful) than those of animals. I feel convinced – though I speak under correction – that it is chiefly due to the excessive assaults we make upon the alimentary canal. I was very struck with the reply once made by a slim attractive ballerina, well advanced into middle age, to the question of how she kept her figure. She said that on one day she would eat meat only; on the next perhaps only cheese; on the next just fruit and vegetables. This brought her closer to animal practice than is usual, for though the animals seldom dine out or think much about a 'balanced diet' they do not punish their stomachs by excess of variety or sheer brutality in content. Consider H. G. Wells's account of Mr Polly's interior which clouded his exterior view of the world, as the story opens. He had just consumed cold pork held over from Sunday, some cold potatoes with mixed pickles which included three gherkins, two onions, a small cauliflower head and several capers. This was followed by cold suet pudding with treacle and then pale hard cheese. The whole was taken with three big slices of greyish baker's bread washed down with a jugful of beer. It is not surprising that wonderful things were going on inside Mr Polly. 'Oh! wonderful things. It must have been like a badly managed industrial city during a period of depression; agitators, acts of violence, strikes, the forces of law and order doing their best, rushings to and fro, upheavals, the Marseillaise, tumbrils, the rumble and the thunder of the tumbrils . . .' It is still less surprising that Mr Polly's view of life was somewhat jaundiced since such a meal was the norm in his house.

It is amazing enough that his canal could cope with it at all. But the sheer capacity of our digestive system to cope with ruthless assault is one of the most striking things about it. Edward VII, we are told, began his day with a glass of milk in bed before coming down to breakfast which consisted of platefuls of bacon and eggs, haddock and chicken, toast and marmalade, with coffee. After one hour out shooting he returned to drink a plate of hot turtle soup. This was followed by a hearty open-air lunch. Coming back for tea he helped himself to poached eggs and preserved ginger, together with scones, hot cakes, cold cakes and Scottish shortcake. This was his preparation for dinner at 8.30 p.m. consisting of twelve courses, with champagne and cognac.

We have not yet tackled the question how the waste matter is separated from the good stuff, and by what means it is evacuated from the system. This is accomplished by the kidneys and the large intestine. The kidneys. Plural: we possess two. As a definition of the kidneys, we might fairly call them clearing stations, even custom houses where what is inadmissible is rejected. Or more briefly still, filter units. Everything in the blood is absorbed by certain tubes; and then, after a selective process, another set of tubes returns the blood, purified, to the circulation. That which is not accepted is discharged into a main tube we call the bladder by means of which we pour out the liquid, now given the term urine, in the manner so familiar to us all.

These organs, the kidneys, are placed in front of the twelfth ribs close to the spine. We would expect them to be large, but they are quite small, hardly weighing as much as 5 oz. The filter units, called nephrons, are so many and so minute that a person has one million and two hundred and fifty thousand in each kidney. The portion of the nephrons which is responsible for the initial extraction of the blood is bulb-like and thus called a glomerulus. It is through the tube connecting each glomerulus to the main ducts of the kidney that the needed substances are returned to the blood system through the renal vein.

Since we cannot absorb all that we swallow, the importance of these clearing stations that we call kidneys is very evident. Without them our blood circulation would become clogged up. It is certainly an advantage having two stations, for if something goes wrong with one of them we can still carry on, though in a precarious condition. It is calculated that seven thousand people in Britain die of kidney failure every year. Still, each year medicine advances in this field, and already a given diet with the aid of a 'dialysis' machine adjusting the blood's composition twice a week can now prolong the patient's life almost indefinitely.

More than two pints of blood are pumped through the kidneys every minute of our lives. This work, of such intricacy, carried on ceaselessly by the massed multitude of the nephrons, is done without our intelligence taking any part in it at all. And when conscious action is called for, if in good health, we experience no difficulty in the evacuation of the urine. Unfortunately the same does not apply when we wish to evacuate faeces. On account of the greater solidity of the product, its exit is not always easy, seldom complete, and sometimes impossible.

In the small intestine all the nourishing matter in our food has been sorted out, and, with the assistance of the lymphatic system, sent into the blood stream. The rejected material passes on into the large intestine becoming increasingly solidified as it goes from the caecum to the colon and from the colon to the rectum in a series of vermicular movements. It is these vermicular movements, or worm-like spasms, that we become conscious of and seek to assist, meeting with relief or disappointment, as the case may be. Indeed, this stage is not only a conscious one for us, but also an anxious moment, because if the faeces get stuck we suffer from constipation – yet today easily overcome by use of roughage in the form of bran or normacol with, if necessary, a pinch of 'cleansing herbs'.

IX The liver

I HAVE NOT yet finished this chapter. Some readers may wish I had, as indeed I do myself. Others would be pained. They would consider it a shame if a certain organ were not mentioned, and ashamed of me for the omission. We might well think that the heart, the lungs, the blood, the kidneys, the vascular system, the lymphatic system, thus working together in concert, stand in no need of further assistance. In this we would be mistaken. They do need help. They need, as it were, an overseer, a foreman, a guardian who will watch out for defects, supply deficiencies, assist the digestion and act as a general dispensary in times of need. This is the liver.

It is much larger than the heart or the kidneys and lies tucked up under the right ribs, containing on its under-surface a small bag known as the gall-bladder. It consists of little bunches of cells called lobules so tightly packed together as to give the liver its well-known compact structure. The lobules are enmeshed in a dense framework of capillaries. This tells us nothing much. If I add that it is often referred to as a chemical factory and is responsible for over five hundred functions in promoting the metabolism of the body, I have only opened a door. I do not propose to go in and inspect that factory and then sing its praises, always supposing that I could possibly grasp the intricacies of the mechanism. I will confine myself merely to stating its role with regard to the digestive system.

It manufactures bile in the gall-bladder and distributes this yellowish fluid to the intestines in order to further assist the

47

enzymic work of digestion. This was once considered a very important function of the liver, but is now played down, so I need not dwell upon it here. However the following definition of the liver with regard to nourishment preserves its validity: 'The liver is the chief seat in the body of the formation of glycogen, a substance like dextrin, which readily undergoes conversion into sugar.' We all need sugar, whether we take it indirectly from vegetables and fruit, or neat from Tate and Lyle. We know that too much of it is bad but also that too little of it is disastrous. We take some sugar (or glucose which is a less concentrated form of sugar) into our bodies, and then according to the best traditions of the digestive system it is transformed and diffused. The wise liver, with a view to a rainy day, contrives to create a kind of sugar bank, by undiffusing the diffused and re-transforming the transformed into a substance called glycogen, and then glycogen into glucose and glucose into sugar. When the tissues need sugar the liver has it ready; when it has too much it is stored up. In short, when the blood is over rich in sugar, as after a meal, it is stored up in the liver as glycogen; and then in the intervals between meals it is sent out again in driblets to the blood as sugar. The liver is equally adept at transforming glycogen into sugar and sugar into glycogen, the latter being the best form in which to preserve and store the stuff.

I have confined myself here entirely to the offices of the liver in relation to the digestive system. I need not stress that for centuries it has been regarded as an organ of the greatest importance in many functions – in fact we cannot exist without it. It has been called 'the seat of life'. The Ancients consulted it for 'propitious signs'. Kings of Babylon 'looked into the liver' as an oracle before making proclamations or issuing imperative demands. In modern times it has attracted less veneration, and with equal absurdity has been associated with biliousness, ill-temper, and pugnacious audacity, and even Hamlet cried out in the most famous of all his apostrophes: 'It cannot be but I am pigeon-livered, and lack gall to make oppression bitter.'

We started this enterprise with figuring a man standing in a field. So far we have given him a heart and lungs and blood; we have given him kidneys and liver; we have given him an elaborate vascular and lymphatic system. Yet he has no right to be standing there.

We Stand Up

❦

I Bones

HE HAS NO RIGHT to be standing there, for he has no means of doing so. He should be but a lump upon the field, no matter how interesting in itself. To stand up he requires a framework of sticks and posts. In fact he requires bones.

Actually we do begin as lumps and cannot stand up, or even crawl for several weeks. As babies our bones, such as they are, give us little support before growth. 'Before growth.' I sometimes marvel how easy it is to utter such a phrase. As if growth were something to be taken for granted. Yet a full grown man should appear a less extraordinary phenomenon than a lump of flesh on the floor possessing the capacity to become that man. Still, I must not pause to brood upon the unobviousness of the obvious, but get on with our story. The Principle of Growth is there, and we can describe it – though not explain it.

What are the bone-builders employed by the Principle? They are cells. We have not thought of cells in this way before. We have seen them in their massed millions composing the substance of our bodies, and as mouths feasting upon the ingredients in the blood-stream to sustain those bodies. Here we see them as bricklayers and builders. The new-born person begins with some bone, but chiefly with cartilage or gristle. There are cells (over a thousand per square inch) that select minerals to make cartilage at each end of the bones, so tiny at first. To take the easiest example, we are born with a thigh-bone 4 inches long. Its upper end fits into the socket of the hip-joint, and its lower into the knee-joint. At each end the cartilage-building cells gradually increase the amount of the gristle. At the same time another set of cells, bone-corpuscles called osteoblasts (again thousands in the space of an inch) invade the cartilage territory with stronger building material. After five years they have added much bone to the bulge of cartilage, usurping some of that space; and they have also gone to work on the shaft, making it thicker and longer. By the fifteenth year it is three times the length it was at birth. The colonies of bone-builders have invaded almost the whole of the cartilaginous area: except at two specific places. This explains how the shaft is made longer. For at each end a layer of cartilage persists, forming a partition, as it were. In these partitions, or discs of cartilage, growth continues, and as the cartilage builders lay down new material the adjoining bone-builders, or osteoblasts, invade and occupy it. These cartilage partitions are constantly growing, and thus the length of the bone is steadily increasing. After some twenty years the cartilage discs have disappeared and the end pieces of bone have become firmly annealed to the shaft, and then there can be no more growth.

It is estimated that about a million bone-builders are engaged in the construction of the thigh-bone of a newly born child, and that on completion an army of one hundred and fifty million has been at work, together with an auxiliary army of cells of greater size whose task it is to absorb degrading tissue and hollow out cavities within the shaft. These cells are known as osteoclasts which thus co-operate with the osteoblasts in the general work of construction.

It is usual to refer to these operations in military terms and speak of invasion and territorial occupation, but it would seem that the apparent discord is really a concord, for just as the osteoblasts act in concert with the osteoclasts, so the final configuration of the bone relies as much upon the resistance of the cells of the gristle as upon the bounty of the bone-makers. However, after the bone has been completed the cells of the former vanish while the cells of the latter stand by ever ready to work at repairs. Thus if we break that bone we need not worry because the osteoblasts in their thousands will soon restore it, if we keep it steady. A repaired part does not mean a patched-up part. Rather it is renewed and strengthened. A man who broke his limbs continuously might claim without impropriety that this was a measure for preserving his youth, since he would be continuously providing his members with renewed fibre; but he would be obliged to watch it as time went on, since age wearies the cells, and the years condemn the bones to brittleness, so that the osteoblasts can perform but perfunctory service. It is all right to break your thigh-bone at thirty-six, but if you break it at ninety-six, as Bernard Shaw did, you are likely to die. But in our prime the sheer power of the osteoblasts can be remarkable, for if a bone be cut out and the membrane surrounding it be left, the bone-builders will endeavour to make, and have on occasion succeeded in making, a new bone. 'I have that man's thigh-bone in my surgical collection at home,' declared Dr Joseph Bell, a famous Edinburgh surgeon, who while going his rounds in the Edinburgh Royal Infirmary pointed to a young man briskly entering the ward.

The number of bones in each person is just over two hundred, and they have all been built in the manner described. Each bone is in a sense two bones: the outer part mineral, the inner organic. The mineral is what we think of first, and it is chiefly made of lime and calcium. The bone-builders select from the blood-vessels precisely those ingredients which make that hard surface that we know as bone. The bricks were fashioned long ago in dateless lands and seas. Indeed, we must rove back through wild centuries of vulcanicity in the ocean beds to come upon the ancestral lime that serves us now so well. Lime is chiefly squashed shell upheaved from mighty depth of waters to form the many mountains whose eroded rock flows in our veins and builds our bones.

All bones have this outer dense compact layer, and an inner meshwork of porous material some of which is composed of marrow nestling in cavities carved out by the auxiliary force of osteoclasts. In fact about 30 per cent of bone is of animal substance which if boiled we call gelatin. This is the substance that is extracted when we make bone soup. Even in fossil bones the animal matter is often retained, and Dr Buckland, of King's, claimed to have made soup from the bones of a prehistoric hyena. It is the porous part that gives to our bones the degree of elasticity which is so necessary. If we fall head-backwards on ice our heads will actually bounce: and if they did not they would be seriously cracked. The curvature of long bones is an aid to elasticity which in the case of our ribs facilitates the expansion and retraction of our lungs; Arab children used to make excellent bows out of the ribs of camels.

After life has left the body the inner layer of bone perishes in a few years, while the outer, losing all elasticity, becomes brittle, a skeleton, the Greek word for dried-up. Not one person in all the world through all the centuries has ever been able to look with calm at the insensate, senseless grin upon the fleshless skull. 'Get you to my lady's chamber,' says Hamlet, 'and tell her, let her paint an inch thick, *to this favour she must come.*' Mankind has never been able to bear this reality and has always sought to evade it, whether like the Egyptians building the Pyramids as a protest, or, as in our own age, burning the body. 'Human kind cannot bear very much reality,' said T. S. Eliot. A much-quoted remark, but I must say I had often thought exactly the same thing long before the poet uttered it. Once, standing with my twin brother on a golf links along with our father, at a tee, I mused: 'To think that we two men, my brother and myself, came from a sperm of this man standing beside us!' It put me off my stroke. Perhaps I should be expected to confront reality a bit better than that, and not be put off my stroke. But I still can't face poor Yorick, any more than the countenance of the corpse: once so close, *now* disinterested, important and alien.

A skeleton suggests framework only, the means by which we can stand erect. But it is much more than that. Indeed, physiologists hold in question whether the office of framework is not secondary to that of reservoir. Of reservoir? It is surprising that we should think of bones in terms of a store house. Yet so it is. I have mentioned that one of the main ingredients to be found in the cavity of our bones is marrow. This is something we all have heard of, and sometimes refer to, saying that such and such an horrific sight 'froze us to the marrow of our bones', just as a certain sort of person will say: 'I feel it in my bones.' Few of us stop to ask what marrow actually is. It is the substance in which the corpuscles of the blood are renewed. Earlier in this book, when dealing exclusively with blood, I mentioned that the corpuscles live for a given time, then die and are replenished. I left it at that, for I could not speak of bone since it did not then touch the main region of my song. Now I can, and I learn that the marrow of our bones provides the chief nesting place for the creation of blood corpuscles. We start our lives with a bag of blood, as it were. We soon need a refill. We need refills all our lives. But where from? From our mutton chop? From our glass of milk? In a sense, yes; when I eat my chop I am adding to my blood, when drinking milk I am drinking blood – though not literally. We have factories in our system where food is turned into blood by virtue of chemical alchemy, and the marrow in the cavities of our bones specially aids this work. With the help of blood we build bones, and within the bones thus made by blood we create more blood to make more bones.

Our bones also provide a permanent mineral reservoir. There is even argument as to whether the structural task of the bones should be considered secondary to 'the more important duty' according to Anthony Smith, 'of providing a mineral reservoir. It is hard contemplating a blancmange of a human being, totally without a rigid framework, but it is harder still imagining the proper function of metabolism without extra supplies of calcium and phosphorous being always readily available.' I was surprised by those words, for I had thought that our food after being transformed into chyme nourished us, with only the liver to check up on deficiencies. It had not occurred to me that this right arm of mine with fist and fingers was not only an instrument with which I could pick up a pen, take a plate from a shelf, or knock down Muhammed Ali, but was also a reservoir of minerals to give strength to that arm. But it does further help me to understand how the gastric and pancreatic juices have something to digest if a person is starved, until at last he seems to be composed almost only of bones, and to look like that man at Belsen pictured on news reels and documentaries, who haunted a whole generation of those who saw him.

In one of Conan Doyle's stories an investigator wished to meet a certain professor, a peppery and difficult man. '*Can he talk beetles?*' asked the professor sternly. On being told that he could the interview was allowed. I can appreciate Mr Venus's claim to Mr Silas Wegg in *Our Mutual Friend*: 'Mr Wegg, if you was brought here loose in a bag to be articulated, I'd name your smallest bones blindfold equally with your largest, as fast as I could pick 'em out, and I'd sort 'em all, and sort your wertebrae, in a manner that would equally surprise and charm you.' But I can sympathize with the girl who turned him down. 'I do not wish,' she said, 'to regard myself, nor yet be regarded, *in that bony light*.' Personally I have no desire to be able to 'talk bones' beyond a certain point. But I do think we should know about our vertebrae. It is perhaps open to question how far we can be said to be 'as old as our arteries'; but it is fairly obvious that we are as old our backbones; and if we are wise we will keep them supple from an early age, that is, at a time when we don't feel the need to do so. Having failed in this, and when in pain, we can still mend matters a good deal by hanging from a bar at intervals of ten seconds and *stretching* the vertebrae, thus preventing them from getting clogged, keeping in place the elastic fibrous discs by which the segments of the column are articulated. On the whole, I think I would be content to leave the rest of my bone knowledge in the bank, that is with the experts, drawing out the knowledge from them as need arises.

II Muscles

MY ATTITUDE is the same towards muscle knowledge. In the 'naming of parts' the bone specialists have announced two hundred and six. Apart from the backbone I am indifferent to these names. The muscle experts, in an ecstasy of nomenclature, have clapped six hundred names on their specimens. I am compelled to be content with finding out just what a muscle is.

Pillared by bones it would seem that we can stand erect. But that is not really possible. It would be an impossible balancing trick without further assistance. 'Station is no rest, but one kinde of motion,' wrote Sir Thomas Browne in 1646, 'that is an extension of the muscles and organs of motion maintaining the body at length or in its proper figure, wherein although it seemed to be immoved is nevertheless not without all motion, for in this position the muscles are sensibly extended and labour to support the body, which permitted unto its proper gravity would suddenly subside and fall unto earth, as it happeneth in sleep, diseases and death; from which occult action and invisible motion of the muscles in station proceed more offensive lassitudes than from ambulation.' In short, while standing up, however stiffly in appearance, we are continually moving and shifting our weight about by virtue of our muscles. Otherwise we would not be able to bear the strain. It is more exhausting to stand than to walk – and for some people, including myself, more exhausting to walk than to run.

In order that we may simply stand upright we are obliged to bring into play no fewer than one hundred muscles. What exactly is muscle? Flesh is muscle and muscle is flesh. When a bundle of it, attached at each end to bone, is *contracted* in obedience to an act of will, the two points to which it is attached come nearer to each other. By this means we can command an almost limitless amount of movement. We can test the truth of this statement at once by putting into action the bundle of muscles we all know best – the biceps.

The muscular system is the second most complex, and in its executions the most beautiful, aspect of the body. Nearly all writers on physiology refer continually to the 'mechanism' of the body, and it is necessary to use the word at times. Yet it is scarcely a permissible term. 'The narrowest hinge on my hand puts to scorn all machinery,' said Walt Whitman. Some physiologists go so far as to compare muscular action with the internal combustion engine, since there *is* a moment of combustion in the cells in the promotion of movement. But I do think that biochemistry is too complex, too mysterious, too occult, to be subject to mechanical analogy. Nevertheless, it is instructive to watch the biochemists pushing their way ever nearer to the discovery of the molecule which is the universal carrier of energy in the cell – when mass becomes energy. Indeed, it has already qualified for the initials ATP, which stand for adenosine triphosphate.

The physiologists speak of voluntary and involuntary muscles. The muscles in action within the alimentary canal and the heart and suchlike perform without our intelligence. But the 'voluntary' muscles, though first asked to act by us, then do so without our intelligence taking any part in the performance. I used to walk regularly from 5 Guilford Street in London to the British Museum Reading Room. But I did not walk there: I was walked there. As I went along I thought of Hoa-Haka-Nana-Ia, the Easter Island sculptured god who then stood at the top of the steps, and whose steadfast gaze steadied my resolve, while all the time my legs took me there as easily as a taxi. When I sign my name I make an act of will, but my fingers move in response to a battery of little 'engines'; the wrist is steadied by another lot placed in the forearm; in turn the forearm is controlled by others in the upper arm, the upper arm by others in the shoulder, while the shoulder is made firm by bringing into play a battalion of 'engines' which support the backbone. No wonder forgery is so difficult.

Just as we should understand the importance of our spines,

we should also understand the relation between joints, ligaments, tendons, sinews. The sinews (or tendons) are really prolongations of the muscles; we might call tendons (or sinews) the *reins* of the muscles. Ligaments are cords of tough tissue which connect bones and hold our joints in place. For example, the thigh-bone (or femur) is connected with a leg-bone (or tibia) by a ligament, and in consequence the knee-joint is able to function smoothly. Failing ligaments, if we moved at all we would move in jerks like robots. If we understand the use of these properties we are less likely to abuse them.

We know that if we oil a sticky hinge it will function smoothly. It becomes obvious that correct exercise is the oil we should apply to our bodies. On more than one occasion I have had reason to go to an osteopath. Years back, he told me to buy and study a certain book. I failed to do so. Visiting him – a man of rare quality – again some years later, he reprimanded me for not getting the book. Shamed, I went at once and got it. The exercises were, of course, based on Yoga. Nothing in the least elaborate. No lotus postures and all that. Just a sequence of simple, if taxing, exercises that brought every muscle into play, *and*, on account of the breathing sequence which accompanied the movements, forced you to stretch your lungs. If only we had the sense to oil our muscles in this way early in life, at the time when we feel no need to do so, we would remain supple into old age.*

In contrast to the plants which prefer to keep their station, the animal kingdom likes to keep moving. The simplest example of this is provided by the earth-worm. Though given time and occasion the worm can bury a city or remove a mountain, it is not notable for the number of its limbs. In fact it has none. But it has two sets of muscles, one running along its body, the other passing around it in rings. When it contracts the former it draws its head and tail towards each

---

*See Rajah of Aundh, *The Ten-point Way to Health*.

other, which shortens its body; when it contracts the rings its diameter is diminished, and in consequence the body is lengthened. By performing these two actions alternately the worms manage to get along. At the other end of the scale in muscular activity, take that prince of animals, the okapi, now so rare that it has to be protected by men from men. That aristocrat of the African wild, with four stomachs, with eyes that can look in different directions simultaneously, with the legs of a zebra, the body of an antelope, the gait of a giraffe, the swiftness of an ostrich, the courage of a tiger, the aloofness of a cat, made such an impression upon the inhabitants of the Ituri Forest that it was held in the same veneration as that bestowed upon deity. This is not surprising, for sheer grace of *movement* always compels attention. We never turn our heads to look at a cow in the field from the train window: but if a deer is seen everyone is excited – because of that grace.

Still, it is to mankind that we must look to see the full possibilities of muscular action. The muscles required by the speaker, the singer, the pianist, the painter, are not deployed by the animals. And when we wish to see in planned executive motion just what the human 'machine' (!) can accomplish, we turn to the ballet dancers and the gymnasts. They are so marvellous as to be almost boring: but when the indefinable quality of Spirit emanates from any of them, the fascinated response of the multitude is very notable. Indeed, motion is also communication: it is language – 'Fie, fie upon her!' cries Ulysses of Cressida,

> There's language in her eye, her cheek, her lip,
> Nay her foot speaks, her wanton spirits look out
> At every joint and motion of her body.

Certainly motion-language was never more emphatic and appalling than with the Fascist goose-step, when human beings pertaining to the condition of robots became fit vehicles by which dictators could vomit their spleen across the shuddering earth.

To return to our first consideration. We alone can employ muscles that permit us to stand on two legs and thus dispense with the third and fourth, turning them into arms. None of the animals can do this. They try, but they cannot keep it up, whether polar bear or monkey, and it is interesting to see how the orang-outang avails himself of a long stick. True, we see the penguins doing better, but in spite of the dignity of their mien their posture is always a failed imitation of Charlie Chaplin.

III Skin

OUR MAN STANDING in the field is now erect and can move about. Yet, as so far presented, he would be an awful sight. All bones and muscles and canals and arteries. He is still unclothed. We have not yet given him the wonderful sheath which encloses and covers all these things. This integument consists of two portions, two layers: the first is superficial and is constantly being shed in the form of powder or scales composed of minute particles of horny matter, and is called the epidermis; the second is called the dermis, which is dense and fibrous. The cells of the former are dead and thus without feeling; the cells of the latter are alive and intensely sensitive. I suppose this could be claimed as a definition of skin. 'Give me your definition of a horse,' demanded Mr Gradgrind of Sissy Jupe in *Hard Times*. The girl was too alarmed to utter a word. 'Girl number twenty unable to define a horse,' proclaimed Mr Gradgrind to the assembled school. But the boy, Bitzer, at once came up with a sharp definition: 'Quadruped. Graminivorous. Forty teeth, namely twenty-four grinders, four eye-teeth and twelve incisive. Sheds coat in the spring; in marshy countries sheds hoofs too. Hoofs hard, but requiring to be shod with iron. Age known by marks in mouth.' Mr Gradgrind was gratified. 'Now girl number twenty you know what a horse is.' But with regard to skin I have not yet done any better than

Bitzer with my integument and epidermis and dermis in holding up the mirror to the marvel of our earthly envelope. In *The Book of Job* Satan is represented as being particularly observant in one respect concerning mankind. 'From going to and fro in the earth, and from walking up and down in it,' he had become hellishly aware of the fact that while a man will bear the loss of property and even children with some degree of fortitude, equanimity, and piety, he will not submit without protest to a disfigured *skin*. 'Touch his bone and flesh,' Satan says to God, 'and he will curse Thee to Thy face.' And, as we know, God took the hint, inflicting Job's body with boils and sores – with good results.

The facts about the skin in relation to our metabolism are important and interesting, and I will mention them in a minute; but the *truth* about skin is the almost mystical fascination which it exerts upon the beholder. There is something about it that frightens us. 'To the natural man, always and everywhere, even today,' wrote Havelock Ellis, 'nakedness has in it a power of divine terror, which ancient men throughout the world crystallised into beautiful rites, so that when a woman unveiled herself it seemed to them that thunderstorms were silenced, and that noxious animals were killed, and that vegetation flourished, and that all the powers of evil were put to flight.' The same writer made a note in October 1912 when sitting on the rocks overlooking a great expanse of sands at low tide. Glancing up from a book he was reading he saw in the far distance a woman in a long coat with three dogs. Half an hour later he chanced to notice the same slender figure walking down to the sea, leaving a little pile of garments in the middle of the completely deserted sands. After her bathe she returned to the pile of clothes guarded by the dogs. A white towel flashed swiftly for a few moments and then 'with amazing celerity the figure had resumed its original appearance and decorously proceeded shorewards.' Havelock Ellis commented: 'In an age when savagery has passed and civilisation has not arrived, it is only by stealth, at rare moments, that the human form may emerge from the prison house of its garments, it is only from afar that the radiance of its beauty – if beauty is still left to it – may faintly flash before us.'

That was in 1912. We know what has happened since then. The Victorians so overweighted themselves with clothes (putting on slightly more when they went to bathe) that the reaction is not surprising. But whatever the attitude of mankind has been in any age towards nudity, the skin has nearly always provoked a desire to decorate it. Sometimes the whole body, or just the face, has been treated as a canvas to be daubed, often hideously, and at other times with extraordinary artistic designs bestowing astonishing effect in personal address or hieratic appeal. And then, not only to decorate it but to add to it, put another skin upon it, clothes. The history of clothes, the philosophy and meaning of the history of clothes, has been shown, especially by the late James Laver, to be far more than skin-deep. Uniform can work wonders. I recall a sergeant-major who, dressed in full regalia, walked as a king amongst men and commanded petrified obedience. One day I met him in mufti coming out of a grocer's shop. All his magic was gone – as if clothes had been the man! The additional attraction bestowed by certain dresses does justify the commercial fuss devoted to this end, though it is hard to forgive the excesses. The most beautiful coat in the world belongs to the leopard: but it faces extinction because there are enough monstrous women who think that they look better in a leopard-skin than a leopard does.

Whatever our philosophy of clothes may be, we should obviously allow our skin to breathe, as it were, and not truss it up too much. It is all too easily degraded, the feet being the most notable victims of distortion. The epidermis registers no feeling, and the native of the tropical forest has so thick an epidermis on the soles of his feet that it is as if he were clad with leather, so little is he bruised by stones or pierced by thorns. This matter is chiefly a question of sane behaviour. We need clothes because they have by now become part of our evolution in a way, these extra skins, often so strikingly attractive. At the same time the skin needs its innings without raised eyebrows or vulgar comment. 'I'll take a shower,' we so often hear a person say. He would be thought eccentric if instead he took his shower in a shower of rain. Yet a shower-bathe, while running, or for that matter a wind-bathe, is as pleasurable as a water-bathe or a sun-bathe. I do think that too many people are too thick-skinned to realize this. Anyway it would seem that the main thing is that the functions of the skin should be recognized and not impeded. As a matter of fact we can't do much to impede its functions even if we deny its pleasures. It is continuously renewing itself. The epidermis, as we have seen, is composed of dead cells (as if we were encased in corpses) which keep flaking off. This is to some extent discernible in our hair. We then call it dandruff, and fondly hope that if we can keep dandruff off we can keep our hair on. I learn from my studies (though I can scarcely believe it) that a large amount of the dust which we continually sweep up is composed of epidermal shavings. It might be wondered how the surface of the skin appears to last so well. It lasts just as the front ranks of soldiers in a battle last – as swiftly as one is knocked out, another from behind takes his place. Incidentally a famous doctor once declared that our bodies are renewed every seven years, and the myth has been repeated by others as seriously as biographers claim (simply because someone once said it without evidence) that Countess Tolstoy copied out *War and Peace* seven times, an assertion as ridiculous as the task was impossible. If it were true that our bodies are renewed every seven years, it is like saying that every seven years our tailor, with extra charge, provides us with an increasingly inferior suit.

The skin's function in excretion is very important. The epidermis is a mass of sweat-glands: two or three thousand to a square inch of skin, and some millions over the whole body, tiny apertures leading to tubes which go with a corkscrew twist down through the epidermal layer of cells to the dermis in a number of loops and coils compacted into a little bunch. In fact it is an extensive and elaborate excretory apparatus. The object of these tubes is to extract some fluid material from the fine blood-vessels which surround them. On an average two or three pints of fluid are excreted by the sweat-glands every twenty-four hours, while many quarts may be passed out by a man working hard in hot weather. The kidneys are not competent to do this job alone. The sweat-glands are essential for the control of temperature, for if it is not controlled fever will be a likely result. But the function of the skin is not confined to temperature control, for the skin is a kind of mackintosh, an impermeable coat helping us to maintain our internal environment which is itself chiefly composed of water. We are really bags of water with things in it. There is the intra-vascular water, and the extra-cellular water, that is, the water between the cells, and the intra-cellular water, that is, the water within the cells. Claude Bernard wrote: 'La fixité du milieu interieur c'est la condition de la vie libre.'

No tennis champion in a five-set match need fear the necessity to go to the lavatory, whereas a company of people after a sojourn in a cold bus will make a stampede for it. Without this excretory network the body would never be able to maintain the 98° of temperature which nature has elected for us as the norm. In a procession to celebrate Pope Leo X's accession to the papal throne a child was covered all over with gold leaf to represent the Golden Age, and within six hours the boy was dead. If this story is true it confirms the fact that the skin is not only a covering but a mantle full of apertures by means of which we can keep cool when we would otherwise be overheated, and preserve the right temperature when assailed by cold.

People make much ado about 'catching cold' or being 'caught in a draught', but the blood-vessels of the skin so speedily contract in response to cold that we need not worry. It is only when we perspire and then stand in a cold wind that the instant chilliness can give us what we call a *chill* – a serious matter. There is a smooth flow of hair over every part of our skin, though not easily discernible, and the phenomenon of 'goose-flesh' is caused by a muscular contraction pulling the hairs into an upright position. But it is not only cold that makes our hair 'stand on end'. Sudden fear can cause it on the scalp, and it is not untrue to speak of 'hair-raising' stories or events: the hair rises on the back of my head if I am suddenly yelped at by dogs. And, as we know, anxiety can cause it to change colour or fall off. During the First World War the colleagues of Lloyd George were astonished at the suddenness with which his hair turned white with the cares of office. An American who was condemned to death lost a thousand hairs a day until he was bald: on the sentence being commuted his hair grew again.

We all know how the skin serves us in temperate climes in relation to the rays of sunshine, shielding us with a pleasant and comforting tan. It seems to me somewhat of a paradox that black should be the colour of the natives of Africa. One would expect them to be very white, since whiteness reflects back the fierce rays of the sun, while black absorbs them. However, unlike the chameleon which can change colour at will, pigmentation is something we can do nothing about, though at one time or another most of us may feel like echoing Donne's complaint:

> Thinke in how poore a prison thou didst lie
> After enabled but to suck and crie!
> Thinke that when grown to most 'twas a poore Inne
> A Province packed up in two yards of skinne.

When we survey someone's perfectly smooth skin we naturally assume that it is just skin, a sheath, an envelope. In this we are deluded. It is also a boarding house. It is the nesting place of as many yeasts, bacteria, viruses and germs as there are people on the surface of the earth; and in addition to these there are often, though invisible, quantities of mites, ticks, fleas, lice and bugs. It will be remembered that the microscope was invented in the seventeenth century. One day in 1674 Anton van Leeuwenhoek when passing a lake decided to take a sample of it for examination through his microscope. He saw creatures a thousand times smaller than the smallest ones he had seen upon a rind of cheese or mould. Then he sampled a pail of rainwater and found creatures 'more than ten thousand times smaller than the ordinary animalcules'. He was delighted to behold structures 'so delightsome and wondrous'; the whole water seemed 'to be alive with multifarious animalcules'. He had discovered bacteria and protozoa. He had opened the door to microbiology and bacteriology. He had laid the foundation for observing the circulation of the blood, and much more.

But when the lens was brought to bear upon the surface of our skins there was less delight at what was revealed. To behold a flea or a mite enormously magnified in Robert Hooke's *Micrographia* and to learn that we frequently play host to such monstrous parasites was alarming indeed. It is not pleasant to contemplate the fact that one's eyelashes are colonized by mites. 'Few people can confront with equanimity,' writes Michael Andrews with justice in his book *The Life That Lives on Man*, 'the idea that worm-like creatures which have been likened to eight-legged crocodiles squirm out their diminutive lives in warm oily lairs in our hair follicles.' This suited the savage satire of Jonathan Swift, for he was not the man to resist an opportunity to depict mankind in a nauseous light. Having hit upon the idea of Big Men and Little Men he makes the Big Man so enormous and the Little Man so small that the latter could roam in the palm of the former's hand as one lost upon a darkling plain. A Brobdingnagian took up a Lilliputian and invited inspection. 'I took him up in my hand and brought him close; which he confessed was at first a very shocking sight. He said he could discover great holes in my skin; that the stumps in my beard were ten times stronger than the bristles of a boar; and my complexion made up of several colours altogether disagreeable.'

The fact is that the microscope reveals the skin not as a smooth surface but full of holes and fissures and ditches and gullies and clefts, with far more hairs than we realize (which applies to women also), while the armpits provide a warm moist jungle for thousands of minute inhabitants. 'I could see the limbs of these vermin,' says a Lilliputian on one occasion, 'and their snouts with which they rooted like swine.' And when, two centuries later, Pasteur related bacteria to certain diseases, it was with little relish that he contemplated the animalcules which had so delighted Leeuwenhoek. He refused ever to shake hands. He carefully wiped any plate or glass before use. He found, or professed to find, in bread served to him, roaches and flour worms.

In days gone by cleanliness was so little regarded as next to godliness that when the martyred bishop, Thomas à Becket, was murdered, his attendants on removing his penitential hair shirt were not surprised to find that it was seething with lice 'like a simmery cauldron'. Naturally there were great plagues from time to time, in which thousands of people died. Today, when there is so much less godliness, there is a great deal more cleanliness. But, as usual, mankind goes too far. Always too far. Michael Andrews very properly makes our flesh creep by his account of the creatures that crawl upon it; but he shows the health risks we run by overdoing hygiene at the expense of ecology. The monumental work by Professor Mary Marples, called *The Ecology of the Human Skin*, had already made clear what we ought to know as a matter of common sense, that if we play host to a large number of minute creatures they cannot all be harmful, some may be necessary, and there must be a degree of symbiosis. We have to tread carefully, for just as the reckless application of insecticides, DDT warfare, has been found – what a surprise! – to be not wholly beneficial, so also the 'conquest' of BO (body odour) by wholesale destruction of protective bacteria is harmful as well as a commercial racket.

We must know the enemies. Amongst many, the flea of

course. We are not drawn towards the flea. Even Signor Bertolotto, who was fond of fleas and a promoter of a flea circus, was perturbed by the behaviour of their larvae. For if two of the little worms are put in a narrow space with nothing to eat, they eat each other. With its mouth each takes hold of the other's tail, thus forming a ring, while they consume one another. When both are satiated they expire. In a larger context we are no friend to fleas, knowing as we do that their plague-carrying power in the Middle Ages killed a quarter of the population of Europe. We are much less plagued by them now, for the vacuum cleaner has largely disposed of them in our homes, although over three thousand can be discovered in a squirrel's nest. It is interesting to learn that when a thousand fleas were removed from a hedgehog it died, for it had become dependent upon its parasites, or possibly because its bloodstream had become so accustomed to the parasitic bloodletting that it killed itself with anti-toxins.

We should not be unduly alarmed by what is revealed by the microscope. Take the *Dermatophagoides pteronissinus*. This fatuous name has been bestowed upon a mite smaller than a speck of dust. When magnified one thousand times it looks like a monster from science fiction, for I fear that magnification is rather a falsification of the real. This mite feeds off the shed skin accumulated in the seams and beneath the buttons of mattresses. If shaken from a mattress these mites can assault the nostrils. If there are too many of them they will be expelled by that pleasurable kind of volcanic action we call a sneeze.

I think many of us are inclined to be ill-informed about bacteria. We tend to regard them as carriers of disease, and it is true that they can be responsible for tuberculosis, diptheria, typhoid and pneumonia; but the conditions are understood and exceptional. For the most part they are beneficent. Their numbers are not to be counted: it would be like trying to count the particles in the clouds. We are covered with bacteria. They do not live long individually: not more than twenty minutes;

but they can multiply so fast as to reproduce themselves a million times in eight hours. It is as impossible as it is foolish to get rid of them by washing, for a hot bath only succeeds in multiplying them on the skin. The smooth face of a pretty girl has millions of bacteria upon it, as if she had multitudinous invisible freckles. A baby starts with none. A day after birth six thousand may be found in a square centimetre of the armpit; by the fifth day there should be twenty-four thousand, and after the ninth day a stabilization of about eighty-one thousand. It is like acquiring a mighty garrison of soldiers to repel attacks by disease-causing organisms, generally called pathogens. Battalions of bacteria in the infinite benignity of their bounty preserve us from harm and turn to life again life's refuse, making incorruptible that which was corrupt. In the promotion of the carbon and nitrogen cycle they have always been necessary. They set up their line aeons before the two first orchard thieves ate those apples; and if somehow they were 'got rid of' there would be no life at all, nothing but the wrecks and shards and memoranda of mortality.

We Feel and We Think

🦑

I The Nerves

YET, EVEN SO, with all these properties of bone, muscles and skin; of lungs and heart; of vascular, lymphatic and digestive systems; of liver and kidneys; our man standing in the field is still in poor shape. He can do nothing. He is incapable of movement. He needs another system before he can do anything at all. He must be able to co-ordinate his organism.

True, I have spoken of his capacity for movement and of his muscles contracting 'by an act of will'. But I had no right to use those words. When writing any book I have a counsel of perfection: give nothing a name before explaining what it is, and make no statement in advance of sequence. I abhor being told in any book, however learned, 'we will deal with this aspect later'. It was as inadmissible for me to say 'by an act of will' at that point, as it would be for a novelist to say 'eventually Naunton Wainright was to strike down the murderous villain with one single blow' and then add in brackets 'for Naunton Wainright see Chapter Eight'. And if anyone wonders why I should be so concerned about form in a book on physiology, I say that it should apply to any book but especially one on physiology. At the same time I cannot afford to make a fetish of my counsel of perfection since that could lead to imperfection in clarity. Our bodies make a *whole*: the unity of the organism, the co-ordination of its parts, is its chief, and most astonishing, characteristic. Consequently the moment we deal with one thing we are obliged to relate it to another, and it is scarcely possible to tell the story of the body in a steadily unfolding sequence. I do not propose to adhere to the attempt too fanatically. Still, I have managed to reach this point without mentioning nerves, and without mentioning the brain. I am now at liberty to approach them.

It is important that any living thing should be able to *feel*. It would be disastrous not to be able to feel pain, since if no such signal were raised at an injury we would soon be in serious trouble. We must have sense. Indeed it is our sensibility that we cherish most and its destruction that we dread most keenly. There is no greater cry of anguish in literature than that of Claudio in *Measure for Measure* when faced with extinction: '... This *sensible warm motion* to become/A kneaded clod...'

By what means then can we be made to feel? Our bodies are composed of all those cells. Are they subject to sensation? If any cell is pricked with a pin, will it give us a shock? (I use that word as the best way of defining sensation.) No, not any cell. Our bodies are cellular communities within which groups are engaged upon separate tasks, all working in harmony to keep the system healthy and alive. One group only deals with sensation: these are the nerve cells.

Each cell is a separate little system, closed in, as the name implies, by a surrounding wall. This wall regulates everything that goes into or out of the cell. The nerve-cells are very long threads, drawn out to make nerve-fibres, cords stretching in an immense network throughout the body, superimposed upon the vascular and muscular systems. In saying that their task is to register the kind of shock that gives the sensation of pain, I was merely mentioning a single office. Their whole task is to keep alert, and awake, and alive, and in tone, and in action, the entire organism.

At first the network was simple enough. We can still see at least one representative of a model of a primitive nervous system in the little creature called the Amphioxus, found in the Mediterranean and even the English Channel. Its nerves are suffused evenly throughout its body, it has in fact a central nervous system but nothing that can be called a brain. As the centuries passed, bones were evolved and the network became more elaborate, and when the backbone came into existence there was a base for a spinal cord of nerves reaching into the skull where an immense *group* of cells, amounting in a human being to about nine billions within a single inch, made what we call the brain. For we must not think of the brain as a majestic organ engaging nerves to act as its minions and slaves, but rather as a necessary creation of the nerves themselves seeking a central station. The nerves served as brains for the very primitive organisms and it was not for many centuries that creatures required a nerve communication centre. The little Amphioxus needed no headquarters, but later on the enormous dinosaurs certainly did need one, and in the case of the Atlantosaurus and of some others two brain-centres were established, one at the head and the other at the rear of the animal. This did not make it cleverer than others, but helped it to be less idiotic.

Let us now get closer to what we mean when we speak of the nervous system and what we mean when we speak of the brain: they are so inseparable that we must consider them together. I ventured to say previously that I was not anxious to be able to 'talk bones' to an exhaustive extent, but that I favoured understanding of the backbone – a somewhat necessary amendment, seeing that the consciousness of mankind depended upon the evolution of vertebrae. That evolution began, it seems, in the Cambrian era which is calculated as two thousand million years ago when the family of echinoderms began to unite, or bridge over to, the family of chordates – in easier terms, when the spineless took on spinehood. We catch sight of the little Amphioxus in Silurian times, that is about three hundred and fifty million years ago. It did not qualify as a vertebrate, for only a gelatinous rod enclosed in a tough sheath with a main nerve running above it took the place of a backbone. Nevertheless this rod (or notochord) served to support the swimming muscles of the small animal and led the way to the beginning of a nervous system when the upper end of the spinal cord developed a nobby ganglion that was the rudiment of brain. Given time, and indeed time was given in abundance, the insatiable evolutionary appetite could work towards the physiology of mankind. And now, without more ado than starting a new paragraph, I will leap across three hundred and fifty million years and come to our own central nervous system.

It has for its scaffolding the backbone, which is a vertebral column, a chain of jointed discs with long arches between them, running from the lumbar region of the back to the base of the skull. The great nerve called the spinal cord threads its way through the vertebral canal behind the discs and rings of the vertebrae from the lumbar region to the brain. This central nerve is connected with every part of the body by means of the branches which lead from the cord, and the branches which in turn lead from them. A nerve is composed of individual fibres, each fibre being part of a cell which *stretches out* and is called an axon (an extension of the basic neuron), while other branches join the axons and are called dentrites. A main branch issues from the vertebrae between each of the rings, at each side of the spinal cord; and when we speak of the nervous system, this is the central part of it. The peripheral system refers to the network – the marvellous meshwork of interconnection – which pervades the entire body and is superimposed upon all the tissues and all the systems. And I use the word meshwork with emphasis because the nerve threads are only one hundredth of a millimetre in diameter, though a given large nerve, such as the sciatic, can be as big as a pencil, being a conglomeration of millions of fibres.

Take an example of a nerve in action. One of my toes is pierced by a thorn. Inevitably a nerve is touched. This instantly triggers off an impulse which very much like a Morse code is carried up to the spinal cord whence it is relayed to the brain. At the same time as the code is ascending to the brain it has triggered off various axons in the lumbar region which compel impulses to descend by another route to the toe, causing it to jerk upwards. The impulse coming from the toe is called afferent, the impulse going to the toe is called efferent and the action is called reflex. Thus most of the process is involuntary or automatic; but when the 'code' reaches the brain an interpretation is made which declares itself, in this instance, in terms of pain, and steps can be taken to avoid or alleviate the sensation. So when my toe is pricked by a thorn or cut by a knife there is a reflex action and an intellectual interpretation of it.

Broadly speaking we can say that every afferent impulse is responded to by an efferent impulse; and we may say that the object of the millions of nerve cells with their axons and dentrites is to make a kind of nerve circulation when every afference is answered by a sequent efference, and the muscles of flexion act when the muscles of extension relax. The greater part of the impulsion and reaction is involuntary; if I bend my elbow on purpose the action excites a multitude of motor centres of which I have no cognition.

I do not think that words are adequate instruments for holding up the mirror to the nervous system, nor do I favour pictorial presentation when too often we are presented with something resembling a horse's tail (at the lumbar region) rather than the actuality of the axionic and dentrical complex which is yet the definition of harmony. We are obliged to use our imagination – which Wordsworth defined as 'Reason in her most exalted mood' – to realize that while we know about this or that reflex, there are thousands, even millions, that we do not know about, since efference answering afference in multitudinous concert throughout the fleshly tabernacle creates the total functions of physical life in reply to stimulation from light or air or heat or cold or odour or sound or touch. It is only the imagination that can brood with any degree of insight upon the play of millions of cells and fibres acting and reacting in co-operative groups to promote, night and day whether we are awake or asleep, the common wealth of our metabolism. We are obliged to contemplate millions of cells coding at the rate of forty to eighty impulses a second year after year: groups in relation to blood-vessels, groups in relation to hearing, groups in relation to seeing, groups in relation to breathing, groups in relation to balancing, groups in relation to smelling.

And still I have not yet said what a nerve is – only attempted to describe its performance. What are these impulses? Why should the cords or wires or threads which we call nerves possess a property not granted to any other thing in the body? At least, what is the property? It is electricity. The impulse which travels afferently towards the centre is electrical, and the cord is akin to an electric wire though the 'message' does not go nearly as swiftly as an ordinary electric current along a copper wire, for it relies upon a series of continuous chemical changes all along the line. However, the short answer is electricity. It is beyond my reach to give a long answer, if long answer there is. But I am not surprised. We are all packets of electricity.

II   The brain

I HAVE JUST USED the words 'groups in relation to balancing'. When I had succeeded in giving our man in the field bones and tendons and sinews and muscles I assumed that he could then move; but of course I was anticipating his system of nerves, for in order to function the muscles must be continuously 'triggered off' by virtue of nerves. Without their aid he could not balance himself upon his two legs, he could not 'keep his balance' as we say. And without the aid of nerves our heads could not be balanced and pivoted upon the topmost vertebra with its peg-and-socket joint. We owe the head and all we have within that house to the co-ordination of backbone and nerves.

The chief property which is housed there is the brain. This organ is about 3 lbs in weight, which seems to me quite heavy if we think of 3 lbs of sugar; we do not feel it at all, and would hardly credit the spongy 'grey matter' with more weight than a few ounces. Its upper structure is in the form of two hemispheres in convoluted folds which are superimposed upon the older tissue. For the brain is unlike any other organ in the body such as the heart or the liver or the lungs. Over the years it has acquired new material sharing space in the skull with the old material, which is often referred to as the old brain in contradistinction to the new brain, the cortex, which is the chief seat of consciousness. The brain possesses some twelve thousand million nerve cells. Colin Blakemore, in his 1976 Reith Lectures, calculated that if the nerve cells and fibres in one human brain were stretched end to end they would reach to the moon and back, for each cubic inch of the cerebral cortex is supposed to contain ten thousand miles of nerve fibres connecting the cells together. I was very puzzled by this at first. Thinking of the extreme *smallness* of the things I thought that a hundred yards in single file would accommodate the lot. But we must not think in terms of dots but in fibre lengths.

Anyway, what we have got here inside the skull is an outcrop of the nervous system that started so simply millions of years ago when a general sensitivity was all that the animal required. Now man needs a central power station. To use a simple figure of speech, by no means qualifying as a metaphor, we might say that the nerve wires in relation to the brain stand rather like the railway lines ending in Waterloo Station with the passengers continually arriving and departing. And, just as at a big station a lot more is attended to than the arrival and departure of passengers, so at the great nerve station which we call the brain a lot more goes on than the interchange of afferent and efferent impulses. J. Z. Young likens it to an enormous ministry with fifteen thousand million clerks all busy coding ten million telegraph wires bringing information to the office. Indeed, I see that Lewis Mumford describes the human brain as serving as 'a seat of government, a court of justice, a parliament, a marketplace, a police station, a telephone exchange, a temple, an art gallery, a library, an observatory, a central filing system, and a computer.'

So, at first then, there was no need for brain; a few nerves would do for the Amphioxus or the Planaria. Yet a pin-point of brain appeared and advanced in other forms of life; the tiny organ gained ground; at first a poor thing; nothing worth it seemed; no more than a fold in the skin; no sign of the potential: yet with power to grow and make for itself a dwelling-place bulging ever bigger like a bubble in what we call the head, and gradually assuming the attributes of sovereignty it made windows in its corridors from which the evolving Spirit could keep watch upon the world.

At this point we come upon something quite new in our story. We find an organ that is not like any other in the body: not like a bone or a muscle or a vein; not like the lungs or the heart, the liver or the kidneys. This organ, the brain, differs in that it embodies an element or principle not to be found elsewhere in the framework. We call it mind. If we eat brains for lunch, just as we might eat a lamb chop, we do not feel that we are lunching on a piece of mind. But mind had been part of it. We do not know what consciousness is. That mystery has not yet been solved. I am in great hopes that it never will be, for I can't see exactly how the cortex could examine the cortex, nor thinking have a look at thinking.

However, we do know that the brain is the seat of intellectual consciousness. This was not always obvious. Aristotle was not intelligent about the seat of intelligence. But he had so powerful a mind that he is regarded as the father of the scientific method, since he was to propose the singular idea that it would be a good thing to observe, dissect, and describe an object before speculating about it. This approach did not appeal to Plato who rejected experiment and observation, and formed the idea that true knowledge can only be attained through pure thought unadulterated by evidence which is always subject to the fleeting falsities of the senses; and it is rather ironic that it was Plato and not Aristotle who deduced the true position of consciousness.

Aristotle thought that the brain was in the nature of a sponge whose business it was to cool the blood from the heart which was the organ of thought and of soul and emotion. It is not strange that the heart was at first chosen, since it beats fast or slow according to our state of mind, and we still speak of a 'heartless' or 'soulless' man if he is lacking in emotion. It is less easy to understand why the liver should have been chosen by both the Sumerians and the Assyrians. 'My liver shall sing praise to thee and not be silent,' sang the Psalmist. In the Bible the word brain is never used, and other organs are given for the seats of thoughts and emotions. We read: 'His bowels yearned with compassion'; 'His veins instruct him in the night seasons'; or 'The Lord trieth the heart and the kidneys.' The ancient Egyptians, with their passion for permanence, entombed the discarded envelope of their emperors with all the props and appointments of life, yet spooned out the brain through the nose lest it corrupt the vessel.

It is strange that the inside of the head should not have been the obvious place to look for mind, since our eyes are there and vision is so emphatically linked with thoughts and emotions. And when we search for a word or the memory of a name we almost feel the effort just behind our foreheads, as well as our sense of personal identity. Also, presumably, the Ancients often hit each other over the head hard enough to cause concussion. The person in question then lost consciousness much more definitely and quickly than if hit anywhere else. They seemed to have deduced nothing from this rather striking fact.

Let us now look closer at the brain. As already noted, the properties of the human organ are peculiar in that it carries compartments which mirror the evolution of species. We still possess the brain as it was when we were in the worm stage of evolution. We call that piece medulla and pons, which occupies a position just above the spinal cord, and still regulates without instructions a myriad of our internal functions such as the action of our lungs and the beating of our

hearts, no matter whether we are asleep or awake, a centre so
vitally important to the working of the body that no surgeon
dares tamper with it. The cerebellum, in the vicinity of pons,
was already with us in our fish stage, and now is concerned
with balance and posture. The hypothalamus, from the
amphibian stage, is in the centre of the brain, and relates to
temperature, appetite and sleepiness. Close to it is the
thalamus – meaning 'hidden chamber' – which is the centre
referred to when we speak of such and such an emotion or
memory or idea as rising up from the unconscious. 'Growing
from the Thalamus but extending over the whole Old Brain
like some crinkled mob cap is the New Brain, the Cerebrum or
Cerebral Cortex, which, from tiny beginnings in the more
evolved amphibians, was enlarged in the mammals and then
in the primates, but has attained its overwhelming size and
importance only in man.'*

Most people would agree, concerning themselves, that the
Me-ness of Me, the I-ness of I, is not to be found and felt only
in the cerebral part of the brain but also in the centres of
emotion that lie below the cortex. But here we come to
something strange and unexpected in the evolution of species.
Hitherto each extension or variation in an organism has
harmonized with the rest of the organism; it was evolved to
fulfil a need, and thus fitted in perfectly and worked like
clockwork. But with the development of the cortex and its four
hundred square inches, we find intense activity not always in
harmony with the rest, the conscious mind not always in tune
with the unconscious mind. A novel state of affairs. And not a
healthy one. If the highly conscious intelligence department is
developed too much and fails to keep in unity with the
unconscious centres, the person as a human being can be
emotionally impoverished. There are people, sometimes
called and not without reason 'brittle intellectuals', who think
and act from the cortex at the expense of the thalamus where

Man on Earth by Jacquetta Hawkes.

it could be claimed that their deepest *being* resides. It was to that centre that Walt Whitman addressed himself in *Leaves of Grass* in the series of poems under the head of Calamus; and to which Herman Melville was referring when he wrote 'far beneath the fantastic towers of man's upper earth, his root of grandeur, his whole awful essence sits in bearded state; an antique buried beneath antiquities, and throned on torsoes!' D. H. Lawrence, too, raged against the over-development of the intellectual consciousness which could lead to loss of touch with the thalamus, which is so intimately connected with the deeper levels of the personality as well as with the sexual organs and the ganglia forming the terminals of the nerves of the eyes and nose and ears. Of course he went to extremes, and to read him one would be led to imagine that the unconscious is rather more conscious than the conscious. But we need prophets who go to extremes, for the path of excess leads to the palace of wisdom, as Blake said in those heavenly counsels which he called Proverbs of Hell.

Indeed the integration of the various departments of the brain is in practice nothing less than the integration of the personality. This was most strikingly brought home by the famous case of Phineas Gage, an engineer who in 1848 suffered an extraordinary accident. Gage, a man of high character and of responsible and warm disposition, was about to blast a rock on a railroad when by mischance the tamping-iron rod which he was using backfired and tore through his skull from beneath his left eye, coming out at the other side of his head. To everyone's astonishment he recovered quite soon. His eyesight remained perfect, his speech unimpaired, his senses intact. Yet a metal rod, $3^1/_2$ feet long, of 13 lbs weight, and with a diameter of an old English penny, had passed through the frontal lobes of his brain taking with it a tooth-brush's worth of grey matter. He had not lost his senses. But it was soon found that he had lost something else. He was deprived of a portion of his *character*. The rod had removed a chunk of his personality. He had been a very pleasant man; he became repellent, ill-tempered, violent, and obscene; and so irresponsible that he could not hold down any job, and was finally obliged to exhibit his head together with the tamping-iron as a fairground attraction. They are still on display at the Medical School of Harvard.

If this event depressed the Cartesians, it pleased the doctors. If a chunk of brain blown out of a man's head was equal to a chunk of character, there was not much room left for the dualism of Descartes; but the unfortunate fate of Phineas Gage focused attention upon the compartments of the brain, and the intimate connection between those departments. The frontal lobes were found to be intimately connected with the thalamus. The lobes of Phineas Gage had been cut off from the deepest centre of his personality: he had undergone a rough kind of surgery. Modern surgeons have now become so acquainted with the intricacies of the brain that they can cure a man suffering from manic depression by snipping off certain nerve fibres connecting the throbbing frontal lobes of the cortex with the old brain. Such a pruning operation can give the patient permanent relief, though it may also rob him of his full humanity.

At the time of Phineas Gage's accident there was not much knowledge regarding the localization of brain centres. But the idea that different mental functions are localized in various parts of the tissue of the brain was beginning to take shape. The pioneer of this idea was a Viennese doctor called Franz Gall. He has been a little unfairly discredited and maligned, not because he was wholly wrong but because he over-played his hand and was the victim of crude followers. He believed that the internal faculties in the brain caused external marks on the skull, such as bumps. He pressed home his case by examining individuals who were especially distinguished by some faculty; and in searching for people with extraordinary heads he naturally found them, and could then draw up a schematic catalogue of the relationship between particular mental characteristics and bumps on the skull, finally bestowing a name upon his theory – phrenology. He produced a model of a cleanly shaven head which was mapped out into *sixty* little plots! He placed the organ of hope on that part of the head which lies over the convolution that actually is connected with the movements of the body; he placed the organ of amativeness in the region of the cerebellum, while that of constructiveness was placed over the inferior frontal convolution. In no case was he correct; but a novel theory, supported by particular instance and given a name, always looks impressive at first, wins attention, and encourages followers. Professional phrenologists, claiming speciality in this kind, came forward and professed to be able to interpret a person's abilities according to the 'bumps', real or imaginary, on the exterior of the cranium. It goes without saying that this 'science' had its vogue, and was especially popular with mothers and fathers anxious about the potentiality of their children. The professors would elaborately map out a head into areas of the crudest sort, the geographical survey being watched by the parents with a credulity only equalled by the solemnity of the phrenologists.

Nevertheless Gall, though discredited, should not be quite written off, for he prepared the way for the great neurologists who, comparing patients who had suffered damage to given parts of the cerebral hemispheres, maintained that control centres, whether auditory, visual, motor, or otherwise, were strictly localized in the brain. Had the rod which pierced the head of Phineas Gage gone through the back of his neck he would have been blinded. His case was of special interest because while his senses were not injured, those elements of the mind such as responsibility and good nature were shot out of his head by the rod. Of course this did not prove that integrity, say, was localized in the frontal lobes. Quite the contrary: it witnessed to the unity rather than the departmentalization the brain; but it stimulated the neurologists to press forward in ever more close examination of brain structure. And it is now understood that while in the cortex there are masses of nerves without precise functions, interacting in such a complex way that full understanding may never be possible, there *are* specific centres with particular functions. For example, the speech control area runs obliquely upwards across the temple towards the back of the head, and is called the Sylvian fissure; there is the visual area which is not near the eyes but at the back of the head; there is the motor centre, controlling muscular movement, which forms a band across the top of the cranium; while smell, taste, and hearing occupy the temporal lobe above the ear. There are twelve specific areas governing twelve specific functions.

Thus behind the smooth forehead of youth or the ribbed and dented brow of age lies this land of many mansions, its components so subtle and intricate, its instruments so ingenious and complex, that to build a 'mechanical brain' on this scale it has been calculated would cost £4,000,000 to construct, and as much to run, and of course without anything in the nature of a mind or holy ghost. 'The latest mechanical calculator in America has 23,000 valves,' writes J. Z. Young, 'but the cortex of the human brain has 15,000,000,000 cells. A computer with so many parts is beyond the dreams of the engineer. A huge building would be needed to house so many valves and all the water of Niagara would not be enough to work and cool them. Yet all that such a machine can do, and much more, goes on gently, gently in every human head, using very little energy and generating hardly any heat.'

It will be easily understood that if an instrument of such delicacy and elaboration in active synthesis is damaged the results may be appalling. If anything goes wrong with the network of capillaries within the brain, it can be a very serious matter. A cerebral haemorrhage means that a blood vessel has burst and caused a clot. According to the size of the clot there will be paralysis to a greater or lesser extent. And according to the area where the injury has occurred, the functions governed there will be affected. The left side of the brain controls the right side of the body, and consequently a haemorrhage on the left side will cause paralysis on the right side, and vice versa. But the effect is not equal; one side is dominant: injury to the right hemisphere has much more serious consequences than to the left. Louis Pasteur suffered haemorrhage on the left side, and yet was able to carry on his work for the rest of his life, though a post-mortem examination showed that he had been functioning with only half a brain – without greater discomfort than a man who having lost the sight of one eye can carry on with the other.

The term 'stroke' is just, for damage to the dominant hemisphere can strike down a person in many terrible ways: paralysis from the waist downwards; or the whole left side, as well as perhaps loss of speech – I mention only a few effects. One of the most tragic and pathetic causes of brain damage is injury at birth. The child then comes into the world handicapped with cerebral palsy. The movements are monstrously distorted and the speech grossly impeded. Not only is this a dreadful thing in itself, but it is not easy to differentiate by diagnosis whether the child is intellectually normal or retarded. Christy Brown, whose motor area is so out of order that he can only write with a foot, and whose speech is extremely difficult to follow, retains his clear mind and literary gifts.* A great deal of additional suffering is caused to the cerebral palsy victim (to be spastic is but one form of palsy) by the incapacity of communication which gives uninformed people the idea that he or she is an idiot.

Complete damage to the area that governs touch is rare. Perhaps of all the five senses that of touch is most important, for if our hands and feet were unaware that they had touched anything, it is hard to conceive what we could do – certainly

---

*See *My Left Foot* by Christy Brown. And also *The Way of Life for the Handicapped Child* by my late wife E. Collis who possessed an extraordinary capacity for diagnosis in this respect, at her unit at Queen Mary's Hospital, Carshalton, where she conducted remedial measures with remarkable success, until she was struck down herself with paralysis on her left side – unrelieved for twelve years before she died.

not build a house, feed ourselves, or cross a street. But if an injury does occur to the parietal lobe, a touch locale, the person is affected in an extraordinary manner. The opposite side of his body becomes *unknown* to him. He does not accept it. In fact he rejects its existence. He fails to see it as part of himself, shaves only one side of his face, combs only half of his hair, and is indignant at being compelled to share a bed with limbs which he does not recognize as his own.

There is another kind of damage which is more prevalent, though still pretty rare. (I will be reproved for saying that, for we do so love to dwell upon our ills.) I refer to the way in which pressure upon the mind in terms of anxiety affects our digestion and gives us ulcers. I used to think it very strange that anxiety had caused in someone an ulcer. It is an entirely material and concrete thing which can be touched. Anxiety is completely abstract, having no material properties of any kind. How then can it become an ulcer? If a surgeon cuts out an ulcer from my body and hands it to me, is he then handing me a slice of anxiety, mercifully removed from my system? Yes, that is so in effect. The frontal lobes include the cortical motor for the unconscious running of our internal organs, embracing the stomach, intestines, and kidneys. The emotion of anxiety causes the motion of secretion, just as sorrow is made manifest in tears; and that secretion hardens and finally declares itself as a corrupt lump of tissue called an ulcer. Insults in the office become poisons in the body. I think that humiliation as well as pressure of anxiety is most likely to lead to the disorder. Your tycoon, however hard he works, is less liable to suffer than his subordinates. 'I do not get ulcers,' one of them recently declared. 'I give them.'

We Have Senses

❧

I / Sight

IT SEEMS THAT our man in the field is now in better shape, as his silent blood glides through the groves and his muscles smoothly move in obedience to the nervous system in its storied structure. Yet his case is still not good. He can neither see nor hear.

To see: a verb of three letters. The eye: a noun also of just three letters – which if unhappily renounced means the loss of a pearl beyond any price. To some small extent we can understand something about this; but, mainly, vision is a mystery even to the greatest of the visionaries.

We see the action of light first in relation to plants. They are sensitive to light; they fed upon it in endless photosynthesis; they were the very children of light; but they could not see and they could not hear; if Solomon in all his glory was not arrayed like one of them, they knew nothing of it, and cared nothing. Then came the animals.

And with them a new idea. The plants had always been, and to this day remain, stationary. Feeding upon earth and feasting upon air they survive without slaughter. It was not necessary for them to see anything or hear anything or pursue anything. But the animals, having no roots in soil, and taking no nourishment from air and light, must move about to seek for food. At first they also could neither see nor hear. But their whole bodies were 'sensitive to light', a phrase which does not mean anything like photosynthesis but simply that the radiation acted upon them. The most primitive action made the creature move towards or away from the light. In the words of J. Z. Young: 'This movement, phototropism, occurs even in the minute creatures, such as amoebae, in which the whole body consists of a single cell, and which therefore have no true sense organs. Some of them swim towards a light, others away from it, and the response is constant for the species.' This was not sight, but was a step towards it, proving that when light falls upon a cell it can make it do something. But when bodies evolved into aggregations of cells those which were sensitive to light were gathered together into a single patch. This concentrated bunch of cells, with the property of being affected by light, was the beginning of eyesight.

Since then all animals have not attained eyesight: the worm does not possess any eyes, and the bat uses them hardly at all. But they are found throughout nature, and are as extraordinary in their abundance as in the variety of their appearance. There are the unblinking eyes of the fishes that shed no tears, for they are always washed (the liquid in the human eye is in the first place for ablution), and there are the compound eyes of the insects. Nearly every insect shows a protuberance on each side of its head which reveals a multitude of six-sided facets symmetrically grouped like pavement tiles. Independent, but assembled in a common cluster, they constitute all together one compound eye. In the June-bug they number about nine thousand on each side of the head, while in certain other insects the number rises to twenty-five thousand. Besides their compound eyes many insects, such as the cicada, the dragon-fly, the wasp, and the bumble-bee, have also simple eyes, the number of three arranged in a triangle on the forehead, where they shine sometimes like little rubies. As for the common fly, one would imagine it to be all eyes by the celerity with which it moves when we attempt to swat it. An eye is a patch of cells sensitive to light, we say. How little that tells about that thing which seems to *live* on the creature or person in the strangest way holding our attention – serving alike for the girl who also speaks (or listens) with her eyes, and for the rattlesnake's

deadly cleft in the glazed blue of the ghastly lens; for eyes
perched on pyramids of bone or waving on the vulnerable
points of the snail's pillars; for eyes brandished on horns or
massed in clusters. I don't suppose anyone any longer believes
in the 'accidental' view of evolution where the chance play of
millions of atoms fortuitously produces phenomena. It is
certainly a bit difficult to think of 'accidental variation' with
regard to eyes. For eyes have not all evolved from one root. As
Bergson pointed out, the cuttle-fish and the vertebrates,
creatures not related to each other, both developed eyes on
their own account in wholly different ways, and from different
parts of the organism. Each took different routes to the same
end. The eye of the bird is adapted to both near and far vision.
The butterfly's eye contains five thousand lenses and fifty
thousand nerves.

If I may use an artless phrase at the moment, the human eye is
a hole in the forehead with a little screen at the far end. Light
forces its way in and gives us a picture. Light is not a substance
like air in which we walk about. We cannot see it as such, for it
does not exist as light until it hits something. This very thing
which causes us to see cannot itself be seen. It consists of
millions of what we call light-waves travelling the ninety-
three million miles from the sun in four minutes. We cannot
see these waves (or rays) but we can feel them – as warmth.
But the sun has not sent us warmth, it has sent us force. When
a boy throws a stone at me and causes me pain it is not accurate
to say that he has thrown a pain at me; the pain is the *result* of
the stone hitting me. Light is the result of certain invisible rays
hitting things and being scattered or thrown back as
luminosity Luminosity is a very great mystery, and there is no
use in our pretending that it is anything else; we do not even
know how much of it is subjective.

If we go into a room and close the door and pull dark blinds
over the windows we will still have plenty of air to breathe;

but light, being a force and not a substance, cannot get in. (It can only go through glass, which is chiefly composed of holes.) Yet if we make one small hole in the dark blind that force will pass through at once. It will do more than that. It will bring in with it a good deal of what is directly outside – a tree, a field, a cow – and we will see them figured on the screen which we have placed in the dark room opposite the hole in the blind. It does not matter how small the hole is, they can all get through with ease; not themselves exactly, but as images of themselves. How so? Because when this entirely invisible radiation coming from the sun hits an object an impression from every part of that object is bounced back into the air, forming part of the strange luminosity. At night the waves of light pass through the dark sky, which is dark because the waves have not hit our portion of the earth, until they strike the moon which then blazes with light. When they come to us during the day they hit everything and bounce off everything as if they were billions of balls. Consider: your television is on in your sitting room. You are watching the programme. But you need not look at your television set to see it. If you have something in the nature of a shining wall on each side of the fireplace, up to the mantelpiece, and it is opposite the television screen, you can turn round and see the picture there. This is because all the time all the rays on every inch of the picture are passing through the room. They cannot be seen but they are there in ceaseless chase across the room, and we can catch them as with a net on that smooth, shiny upright surface beside the fireplace.

If it should seem curious that a tiny hole will admit an outside scene to pass through it, as it were, anyone can test it with an experiment lasting about a quarter of an hour. Take an empty matchbox. Remove the tray and fasten some white paper over the empty side of it. Then make a pin-hole through the middle of the tray which of course is opposite the white paper. Go into a dark room, light a candle, hold the tray in front of it and there on the white piece of paper, the 'screen', will be a picture of the candlelight. No problem about getting through the pin-point of a hole, for of course the object itself sends the rays through; the top ray reaching the bottom of the screen and the bottom ray reaching the top of the screen so the candle is upside down.

Let us now return to that dark room earlier spoken of; for if we do so it will be as if we were going inside an eye and having a look round. Thus, to recapitulate: we shut the door, pull down the blinds (so-called because they make us blind to light), and place a white sheet opposite them. Rude as this equipment is, it is all we need in order to sit inside an eye. Having made a little hole in the blind we see what happens. Light, because it is not something stationary, because it is the arrival, the continuous arrival of force in the form of radiation, passes through the little hole in the blind; and if outside the window there are a tree, a cottage, a pond, a field, a man, all scattering outward the rays of the sun that have come inward onto them, then from every part of those objects rays will pass through the hole and print themselves upon the screen and we have, if the focus is right, an image of the scene outside. We are sitting in a camera, however crude. We are sitting inside an eye, however rudimentary. For room read dark wall of camera; for screen read photographic plate; for blind read shutter; for hole read lens. Then for wall of camera read wall of eye; for photographic plate read retina; for shutter read iris; for lens read eyeball.

Yet this brief, bald statement does not give us an eye that can see. It is less efficient than the camera – for the latter has a photographer to work it. The eye alone is as useless as a camera without a photographer. We can take an incubated hen's egg and detach the tiny portion that should eventually become an eye. If given the correct amount of salts and foodstuffs it can be kept growing in a glass container. There it is, a little bud, as it were, growing, the appropriate light-sensitive chemicals steadily developing until it becomes a recognizable eye – 'alive', yet not attached to the chicken. We can look at it but it cannot look back at us. This eye has no eyesight, for it is not attached to the brain of the chicken. In spite of its intricate structure it could no more see anything than a glass bead.

So it is with our own eyes. We do not see with them. They are the means *by which* the brain interprets what is imprinted on them by the waves of light. What has occurred at the eye-gates is conveyed by the optic nerve to the back of the neck where resides the vision-department in the cortex. (A small wound at the back of the head can cause total blindness.) After the interpretation our eyes become the instruments by which we see. Euclid thought that light streamed out *from* the eyes in every direction.

How is the brain informed with what has occurred at the millions of 'receptors' in the eye-structure? The retina (at the back of the structure) is the final destination of the waves of light. This layer consists of a multitude of cells called rods and cones. In these receptor cells there are about one hundred and twenty million rods and seven million cones which pass 'information' to the brain through the optic nerve which is a stalk consisting of some eight hundred thousand nerve fibres. To be more explicit, the retina is really a portion of the brain substance, albeit attached by the optic nerve to the visual compartment. Its tissue, with its millions of cells, consists of no less than twelve layers, the deepest of which is known as the layer of rods and cones. As for the area in the cortex which interprets the electrical impulses running through the optic nerve, it is itself divided into compartments, one for the interpretation of form, one for movement, one for colour, and so on.

In my studies I have not encountered anything more astonishing or exciting than the structure of the eye, and yet can only give it an inadequate and flat description! I think we need a model of one in order to grasp the full wonder of it. And even then of course a most striking aspect will be absent. Mirrored in the eye we appear to discern the essence of a person's character, his whole being. How much love we can see there, how much hate; what melancholy and what gladness; such fear and such steadfastness; and that cold eye of indifference. Perhaps all of these discernible in the same person from time to time, and much else besides. It is said that anyone who had incurred the displeasure of the Roman Emperor, Caligula, on being brought into his presence, could tell by a single glance at that index whether or not he would be led to execution. And in accounting for the election of Jimmy Carter as President of the United States, the International Academy of Hypnotists, has, surprisingly, named his 'hypnotic eyes' as a main cause.

I have said that we do not see with our eyes but by means of our eyes. But we could not see even then were it not for *memory*. Clear vision is chiefly memory. Without it we would see the world but we would be unable to distinguish things properly. All would be an unpleasant blur, and the face of your friend a nasty tangle of nose, ear and chin. On the rare occasions when a person has been born blind, and has later in life by virtue of an operation gained eyesight, he has often not been happy at first. Some have even asked that the bandages be replaced, others have been sick. It takes them not days or weeks or even months to be able to see properly, but years. Memory is experience, and it is only by experience that we learn to distinguish a cow from a tree, or the pastures from the clouds. Born with sight we learn gradually to sort things out with our eyes just as we learn to walk with our legs and speak with our tongues. The instant correlation of eyesight with experience-memory is a marvel, and we are quite unaware of it.* The man who was born blind and gains sight has memory, yes, but it is

---

*See *The Science of Seeing* by Ida Mann and A. Pirie.

connected with touch. On gaining his sight he can report nothing but a spinning mass of lights and colours. Show him an orange and ask him its shape, he will have no idea until he touches it: 'Let me touch it and I will tell.' Informed that it is a round object, he will say at last: 'Oh yes, I can *see* now how it feels.' Then show him a square object and a triangular one, and he will at first think that they also are round. For months he will be unable to distinguish the simplest shapes, and may become discouraged and unwilling to learn. But if he perseveres he will be able after some weeks of practice to name simple objects by sight, but they must be exhibited at the same angle and be of the same colour. 'One man having learned to name an egg,' writes J. Z. Young, 'a potato, and a cube of sugar when he saw them, could not do it when they were put in a yellow light. The lump of sugar was named when on the table but not when hung in the air with a thread. However, such people can gradually learn; if sufficiently encouraged they may after some years develop a visual life and be able even to read.'

It is difficult for those of us who are born with sight to realize that at first we *learn* to see. We should certainly continue to improve our capacity to see all our lives. 'I am impressed with the fact that the greatest thing a human soul ever does is to see something, and tell what it saw in a plain way,' said Emerson. 'Hundreds of people can talk for one who can see. To see clearly is poetry, philosophy, and religion all in one.' Perhaps the most simple form of seeing was exemplified by Dr Heinroth's goose. This celebrated German scientist had a young goose freshly hatched, whose first sight of anything in this world was the doctor. The imprint on the brain of the gosling was so strong that the creature formed the impression that Dr Heinroth was a goose, and following him around everywhere, would have nothing to do with its own kind. At the other end of the scale we might fairly take Turner, and note what *he* could see, and what he exhibited to the world. One day a woman, unlearned in seeing, and with all the insolence of ignorance, said to Turner: 'I don't see clouds and water like that.' To which he replied: 'Don't you wish you could, Ma'am?' We may deplore the lady, but we are all rather like that: we do not see enough, and we do not realize the extent to which we create our own world with our eyes. The eye of a Red Indian was so keen that he could find the trace of his enemy or his prey even in the mark of a trodden leaf; but a colonist called William Catlin, a painter, said he was once in great danger from having painted a portrait with the face in half-light, which the untutored observers imagined and affirmed to be the painting of only half a face.

Indeed, it is instructive to realize how often the eye is no guide to the reality of objective phenomena. It is not always sensible to trust sensation. We so often see what is not there. On a late winter afternoon I look up and see that in the branches of a clump of fir-trees a big red balloon has got stuck. My eyes tell me this, but I must tell my eyes that there is no balloon caught in the trees, and that instead of the red ball being about five hundred yards away, it is ninety-three million miles distant. In the last century, before it was understood how hot air can twist the waves of light, it was easy to fall into delusion and snare. The eye of the traveller would find objects displaced: that which was above was seen below; that which was beyond the horizon was clearly discerned. When a land surface such as a desert is heated to the extent which results in excessive refraction, then the blue sky above is laid upon the land below and looks like water. We call it a mirage, and this phantom water generally appears when travellers have most desired to find it. The mockery of this cruel illusion is increased by the weird fact that if there are any trees or mountains beyond the supposed lake, they will actually be reflected in this water which is not there. In days gone by, before this was understood, caravans were often found in the desert full of the skeletons of those who, trusting to the testimony of their eyes, had recklessly finished their supply of water.

Indeed, the human eye is not fitted to cope with the vagaries of evaporation, especially when great refraction is caused under a very hot sun. It is then that mariners on a ship can see other ships beyond the horizon. They can see them because those ships are hooked up into the sky in accordance with the phenomenon known as *looming*. The celebrated navigator, William Scoresby, recorded how he saw a vessel far beyond the horizon resting in the air high above the water, and upside down. Sometimes he would see several images of this one ship together with multiple images of other ships so that the distant sky was filled with an armada of air-borne vessels. And sometimes on the coast he would see great cities where no cities were. He would discern streets, monuments and churches; unfolded before his eyes would appear castles and obelisks, towering turrets, cloud-acquainted minarets and spires – none of which existed. They appeared perfectly distinct, he insisted, 'even in the substance of the most uncommon phantasms, though examined with a powerful telescope, and every object seemed to possess every possible stability. I never observed a phenomenon so varied or so amusing.'

These optical illusions are objective, and, indeed, amusing in the sense that they make us muse in a pleasing way. Subjective illusions, or hallucinations, are more bewildering for they take many forms and arise from many causes, whether from emotional tension or drugs or direct injury to the brain. A man who had taken some lysergic acid, according to Dr Russell Brain, saw that the ordinary little star-shapes on his lampshade were transformed. 'These stars began to be filled with colour; they *lived* with colour. One star, I now saw, was a very small (and wholly attractive) *turtle* on its back, its body a maze of distinct colours.' The stars then turned from turtles into highwaymen with pistols, living and moving in a firmament of illumined paper. Next he saw that the walls were covered with very beautiful patterns, continuously changing. 'More colour. Indescribable colour. And all the colours, all the patterns, *were in the wall* in any case – only we don't usually see them, for we haven't eyes to.' Then he saw that his pyjamas lying on the bed 'were *edged with flame*; a narrow, flickering, shifting nimbus, incredibly beautiful, which it filled me with delight to watch. Clear flame; golden-scarlet. Then I understood that this flame *was music*, that I was *seeing sound*.'

The taking of mescaline can have more bizarre effects than curious perception. A man may think that his foot has been taken off or his head turned 180°. 'My feet turned spirals and scrolls, my jaw was like a hook, and my chest seemed to melt away.' Another man declared that his right arm 'is a street with a group of toy soldiers', while his left arm 'goes across the street like a bridge, and carries a railway.'*

These people who took a dose of mescaline were probably not in very good health at the time. But it by no means follows that a person in sound health, and with a strong mind, cannot take mescaline and thereby widen his understanding of perception. Aldous Huxley made that perfectly clear in his lively and delightful book called *The Doors of Perception*. W. B. Yeats, in excellent health, took a dose and at once became fascinated by an advertisement for Bovril on the Thames Embankment, and beheld gorgeous dragons puffing out their breath in front of them in lines of steam on which white balls were balanced. And fifty years before Huxley wrote that book, Havelock Ellis undertook an experiment with the drug when living in the Temple in 1898. He wrote an account of his experience in *The Contemporary Review* in January that year, and called it 'A New Artificial Paradise'. He summarized the effects as revealing an optical fairyland where all the senses now and again join the play, while the mind itself remains a self-possessed spectator. 'I would see thick glorious fields of jewels, solitary or clustered, sometimes brilliant and sparkling, sometimes with a dull rich glow. Then they would spring into flower-like shapes beneath my gaze, and then seem to turn into gorgeous butterfly forms or endless folds of

*A Drug-taker's Notes, R. H. Ward.

glistening, iridescent, fibrous wings of wonderful insects.' Once he was surprised to find that it was raining gold. This orgy of vision was mostly internal, but he also saw outward transformations. 'The gas-jet, an ordinary flickering burner, seemed to burn with great brilliance, sending out waves of light, which expanded and contracted in an enormously exaggerated manner.' The whole room became vivid and beautiful. It was like the difference between a picture and the actual room, a picture by Monet. He remained calm and collected amidst the sensory turmoil around him. It cast a halo of beauty around the simplest and commonest things. And Ellis makes the curious observation, of interest to Wordsworthians: 'Not only the general attitude of Wordsworth but many of his most memorable words and phrases cannot – one is tempted to say – be appreciated in their full significance by one who has never been under the influence of mescaline.' He adds that a measure of robust health was required, for otherwise it could be injurious; but that for a healthy person to try it once or twice was not only an 'unforgettable delight but an educational influence of no mean value'.

The educational value he was thinking of was doubtless the light it casts upon the extremely complex nature of perception. For we all know how much more there is in the world than 'meets the eye'. How very much more we can only realize if we look through an electromicroscope and see, for example, what a leaf is actually composed of. It is then not the leaf we thought we knew: what green groves are these in this illumined land! What tapestries are here! What fair fields, and caverns, and cliffs, and fissures, and unexpected colours, and glad sequestered glades! There are more things to be seen on earth than we can see, just as there are more vibrations to be heard than we can hear.

To see things in a strange light does not necessarily argue a deranged mind, and quite simple conditions may cause untoward effects. Thus, normal individuals, if exposed to excessively flickering lights, may become the victims of uncommon illusions and emotions. According to Dr Grey Walter the sense of time may be lost or disturbed. A person may think that he is pushed sideways in time, that yesterday is at the side instead of behind, and tomorrow is 'off the port bow'. Had the cinema always qualified for the term 'the flicks', and had its popularity not been usurped by television, the effects might well have been disconcerting.

Sometimes the sense of touch may be included with the sense of sight when a person, in an abnormal state of financial anxiety, sees coins which he has only to pick up. Sir James Fowler recounts how at a general hospital in London he saw a man sitting in bed busily engaged in picking up imaginary sovereigns with his right hand and putting them into a purse in his left hand. Though they did not exist he showed much satisfaction at their colour, shape, size, and weight. The supply was apparently unlimited and he continued to accumulate wealth for some hours. 'The areas of the brain in which are stored memories such as are necessary for such conclusions, are being stimulated by messages coming from within and having their origin in nerve cells temporarily functioning in an abnormal manner, and are erroneously concluded to correspond with the realities of saner moments.'* And under the stress of great emotion an hallucination may be present to one sense but not to another. 'Is this a dagger which I see before me,/The handle toward my hand?' asked Macbeth. He tried to clutch it, but it evaded him:

> I have thee not, and yet I see thee still.
> Art thou not, fatal vision, sensible
> To feeling as to sight? or art thou but
> A dagger of the mind, a false creation
> Proceeding from the heat-oppressèd brain?

Dr Fowler, on describing that man sitting in bed helping himself to invisible coins, commented that 'to see a man in

---

*The Sthenics, Sir James R. Fowler.

that condition, a ghost presents no difficulty.' Indeed, the most elaborate hallucinatory experiences are those of apparitions. 'The hallucinated object or person purports to be a physical object,' writes Dr R. J. Smythies. 'And an apparition usually looks solid, throws a proper shadow, gets smaller as it moves away from the observer, moves around the room with respect of the furniture, and may speak to the observer or even touch him.' He adds that in many cases the apparition 'has been seen by more than one observer at the same time – i.e. there are collective hallucinations.'* When an apparition is seen to open and pass through a door known to be locked, this is because 'these abnormal experiences,' according to Dr Russell Brain, 'are often associated with a modification of normal perception such that the abnormal appearance is integrated into the subject's perception of his environment.'† It will be readily appreciated by the reader that it can be no part of my programme to involve myself in the problem of psychical phenomena, in which in any case I would be certain to flounder, though at the same time I am unwilling to sit on the fence to the extent of evading the question: Do you believe in ghosts? To which I am bound to reply that I believe that people quite often *see* ghosts. I am biased in favour of natural explanations for supposedly supernatural things; I harbour a strong wish-nonfulfilment regarding the existence of ghosts, since that would argue that I could become a disembodied spirit myself – a forbidding prospect. A man whom I much admired, Norman Gale, poet and scholar, who once taught me Greek, told me that on entering a room one day he saw *himself* sitting in a chair. This cheered me up. I knew some people at Cranbourne in Dorset who frequently witnessed a coach-and-four clattering down the high street at dawn. This was also a comforting thought, since few coaches have 'a life after death'

*A Logical and Cultural Analysis of Hallucinatory Sense Experience, R. J. Smythies.
†The Nature of Experience, Russell Brain.

– a curious phrase, come to think of it.

We should perhaps include here one other illusion respecting eyesight, namely the illusion of being blind when the subject is not blind: a condition with which many psychiatrists are familiar. The following is a case in point told to me by one. A soldier, because of his exceptionally small stature, always desired to prove himself particularly brave. He volunteered as a sniper. The mortar fire all around him became almost unbearable, yet he would not crawl back. Then suddenly he became blind and so *had* to go back. The blindness was purely psychological illusion. After due preparation, a psychiatrist, under hypnosis, handed him back his sight.

One of the wisest men who ever lived was the sage of Ephesus of whom it has been claimed that 'his every word bears the stamp of genius'. And though he lived five hundred years before Christ, Heraclitus foresaw at that distance many concepts of modern science, including the perception of the subjective factor in perception. But he was too wise to carry it to absurdity; and I feel confident that he would not have insisted upon the complete subjectivity of colour, in spite of the fact that animals are blind to it. We have been taught today that there are no colours which actually belong to any object, but that each observed colour was four minutes ago in the sun. The colours belong to the light waves which when fanned out, as it were, into what we call the spectrum (which means 'appearance') are seen as colours, a good example being the colours of the rainbow. When the waves strike an object a certain number of them are absorbed by it, the rest are reflected away to our eyes, and our brains make the appropriate interpretation of colour. A totally red rose has absorbed all the other waves *except* red which it rejects. When this fact was first established, not a few people, including poets, were distressed. It does not make me sad, it only adds to the mystery of things. There is now a further school of thought which holds that not only does colour not belong to the rose, but that it does not even belong to the sun, and that the whole trick is performed by the brain alone. Even if this were true, I personally don't mind much; for all of us have the same finished article, have we not? If we are all illuded in this way, we might fairly claim that the illusion is itself an illusion.

Anyway, the thought which must surely be uppermost in our minds when we think of our gift of sight is what colour means to us. John Ruskin referred to the 'nobleness and sacredness of colour'. He declared that 'of all gifts bestowed upon man, colour is the holiest, the most divine, the most solemn.' It pained him to hear people speak of it as a subordinate beauty. 'If the speakers would only take the pains to imagine,' he wrote in *The Stones of Venice*, 'what the world and their own existence would become, if the blue were taken from the sky, and the gold from the sunshine, and the verdure from the leaves, and the crimson from the blood which is the life of man, the flush from the cheek, the darkness from the eye, the radiance from the hair, – if they could but see, for an instant, white human creatures living in a white world, – they would soon feel what they owe to colour.'

And if 'the nobleness and sacredness of colour' is in some measure due to our own power to harmonize a stream of impressions into a picture, it in no way diminishes the vision of glory.

II Hearing

OUR MAN in the field can now see. But as yet he can hear nothing. So let us give him ears. As a preliminary statement we might note that the middle ear is composed of three small bones, two of which are called the hammer and the anvil. Their evolution is interesting. The hammer was once part of the early mammals' lower jaw; the anvil was the bone on the base of the skull with which it articulated. During the evolution of mastication with molar teeth a new joint was formed in the lower jaw; and these two bones were taken into the service of the ear. And so with that hammer and with that anvil we have a base on which to shape the waves of sound.

The waves of sound: let us consider them, prior to further detail. When we talk of the five senses we are referring strictly to ourselves. Insects, according to W. H. Hudson, may possess anything between seven and seventeen senses since they 'appear to be affected by vibrations that do not touch us. We exist, it has been said, in a bath of vibrations; so do all living things; but in our case the parts by which they enter are few.' By parts he is referring to paths or gateways. We employ just the ear alone, and I'm sure it will be agreed that we receive quite enough vibrations. In a cubic inch of air there are said to be four hundred million billion molecules. What we call noise is the vibration of molecules, and there is generally something to cause them to vibrate. Just as light travels in waves so does sound. In comparison with the speed of light, which is 186,000 miles a second, sound travels rather slowly, achieving only one-fifth of a mile per second. But whereas light slows down when it encounters the atmosphere, still more when it strikes water, more yet when it encounters glass, and is stopped altogether by most other substances, the waves of sound go faster when the material becomes denser. In water it goes at one mile a second. In iron it speeds up to three miles.

Ordinary sound waves cannot work in a vacuum. They need molecules to vibrate. When we speak of waves we think of water, and we visualize the ocean waves breaking upon the shore. But in this sense the term 'waves' is as much a misnomer as when applied to light. The so-called waves are not moving any more than corn in a field when the wind blows. An enchanting sight is it not? The wind blows strongly, and looking at the field from a certain distance and at a given angle we will apparently see a succession of waves not made of water but of corn. The more intensely that we gaze the more fluently they flow, whether green in June or golden in the August gleams. The wind:

Rustling the oats with soft incessant hiss;
Racing through barley, ripple on ripple away;
Stiffly swirling the wheat with rough Van Goghish sway*

The gale blows and the stalks bend over and then back, but never move from their station. The perpetual bending gives the effect of waves. So with sound. There is no flow of molecules but of vibrations passed along from one group to the next. Throw a stone into a pond. This will cause ripples in all directions; but the ripples are not a flow of water, but rather molecular agitations moving across without carrying the water with them.

The term wave-length does not mean length of wave. One can think of a very long water-wave from, say, Lyme Regis to Portland Bill (were that possible) or a short one of a few yards. But that is not what is meant. What is actually signified is the distance between waves in front from the waves behind – in short, the size of the trough between the two. We are supposed to envisage a number of different sized troughs, the big gaps causing the waves to travel less swiftly; they arrive less frequently than when the troughs are small, so we refer to their *frequency*. The sheer amount and variety of sound waves transmitting in every direction simultaneously is difficult to grasp. They may go to and fro among each other without hindrance or reinforce each other in the most subtle ways.

We catch some of these waves – or vibrations – in the net we call the ear. They beat upon it, and some pass in through the gateway which has three portals: an outer, a middle, and an inner one. The first, the pinna, leads the waves to the auditory chamber called the canal, which in turn leads to the eardrum, a membrane stretched tightly across the space separating the outer and middle parts of the ear. The inner part is encased in a spiralling bony structure resembling a snail's shell and thus called the cochlea, which contains fluid. The middle ear, acting

---

Letters to Malaya, Vol. III, Martyn Skinner.

in a hydraulic capacity in combination with the aforementioned bones called the hammer and the anvil, plus a third bone called the stirrup, *multiplies* the pressure of the vibrations. The fluid in the cochlea transmits the pressure from the footplate of the stirrup to a small terminus called the organ of Corti (after the Italian who discovered it), when the sound waves reach the auditory nerves.

When we speak we send out waves which are then taken to the ear portals and interpreted as sound: the mouth and tongue speak and the ear hears. Yet in order to hear sound made by ourselves we need not speak or send out any waves. It is unnecessary for the stimuli to come from outside. If with tightly closed lips we hum we are making sound with our own vocal cords through direct conduction to the inner ear via the jawbone. If we press our fingers into each ear the sound is greatly intensified, an effect which is apparently due to the conduction of the jaw bone. For this reason when we hear ourselves speak our voices seem deeper than they do when they are recorded – to our surprise and annoyance.

The complexity of the ear-construction is emphasized by the fact that the inner department serves also our sense of balance. Most of our senses are interwoven and act in concert. Certainly, on the occasions when I have had wax in my ears and have put oil drops into them prior to the expulsion of the wax, I have found it less easy to *see*. I cross a street with much more circumspection than usual, for my awareness of traffic is less acute. And when after crossing the Atlantic by air I am deposited at the airport, the effect of the final descent is so great that I cannot hear myself speak, and I discern the port with uncertain gaze and walk with infirm step.

It is worth attempting to acquaint oneself even with the outline of the description of an ear in order to gain some faint idea of the apparatus that fills our world with such wonder and surprise and the music of so many spheres. 'Arise, Monsieur le Comte!' said his valet each morning to the Comte de Montesquieux. 'You have great things to accomplish today.'

We should arise every morning with thankfulness for the gifts of breathing, of sight, and of hearing, so that a great thing may be accomplished – the joy of living. And how easy to injure the ear by too much *noise*, the vilest sin of our times. 'Please turn it down a bit,' said a mother to her daughter the other day. 'Don't you realize, Mother,' replied the girl severely, 'we teenagers *need* noise.' She did not know what hurt she was inflicting upon her mother. She did not know what injury she was inflicting upon herself. She was unaware of how thousands of factory workers are subject to erosion of hearing by the clatter of the 'plant', and how ceaseless noise at a certain pitch can actually affect the brain and make us into morons. There is no voice so beguiling as the voice of silence. It has been abandoned in our day. We hear it not. A few years ago we heard it. Once a year for two minutes. At 11 o'clock on the eleventh day of the eleventh month of the year we heard it in England. It was at first an offshoot of the wars; but it became far more than that; it was raised above a mere ritual to the memory of the dead or a pious prayer in the name of peace; it was raised above the roar into a universal aspiration for the redemption of wickedness and the quest for concord that never dies. At 11 o'clock for two minutes all the work of the nation stopped, its traffic stilled. Suddenly, amidst the turmoil there was perfect silence, there was perfect peace. For thousands of people it was a great spiritual event: for a few the only spiritual event of the year. But men with evil tongues from Church and State assailed it with their satanic power, and exiled Silence for ever from our streets.

We must not think of our man in the field as being subject only to the ordinary waves of sound. He is a modern man, and his ears are attuned even to people talking on the moon.

There are the ordinary waves which come to us when someone speaks or an orchestra plays, and our brains interpret them into specific sounds. They come to us 'horsed on the sightless couriers of the air', the molecules in their massed millions. But we could not speak to each other above the atmosphere by means of waves dependent upon molecules. For this we must use the forces of electromagnetism, when radio waves are enlisted in our service. They are akin to pure energy, which is initially not caused by something (as vibrations are caused in the atmosphere by a thunderclap) but are themselves cause as well as effect. How far physicists can determine their pristine cause I cannot say, for my mind is too blunt to penetrate into this area; but these physicists, or miracle-workers as they seem to me, can and do harness radio waves to such effect that not only can we speak to one another over long distances by converting our voice-sounds into an electrical current running along a wire and then converting it back into speech at the other end; not only that, but *without wire*, sound can be conducted over the whole earth, and, indifferent to vacuum, take wings upon the void so that we may hold converse with men upon the moon. And further-more, just as these waves are undeterred by vacuity they are unimpeded by substance, and can pass through walls into dungeons, or bring messages of succour to people buried deep in the debris of an earthquake, as, recently, to that woman in Rumania who, clutching her transistor, heard the voices of her rescuers.

I spend most of my life taking most things for granted, only occasionally being struck by the enormity of my ignorance and incuriosity about this and that. (At this very moment I don't know what ink actually *is*! And what do I know about india-rubber?) But there are exceptions. I do not take radio for granted. I do not, like a child, take television for granted. Indeed, unless a TV programme is presenting someone attempting to be funny (which is generally no laughing matter) I actually find it difficult to switch off because of the extraordinary fact that people are *moving* in the picture. I am amazed to see a policeman in Bond Street speaking to another in Scotland Yard. I cannot take for granted the fact that a TV interviewer can look up at a screen and talk to a person who, though fully visible, is not there – and may be in China or Peru. On occasion, when leaving Broadcasting House after recording a talk, I have thought, as I went down Oxford Street, how strange that one can go to a place and deposit one's voice, and leave it there like a suitcase in a cloak-room to be taken out at any time from the Voice Library of the BBC. I am still not attuned to the fact that my room is full of unheard voices from England or Ireland or America or France or Germany or Italy, and that I have only to raise my hand to bid them speak. The most frightful aeroplane calamity has just occurred at Tenerife. Two enormous aeroplanes collided on a runway in a Spanish airport. Few passengers survived; the wreckage went up in flames; nothing could be salvaged; nothing remained in the ruins – except the instructions from the control tower! A little *black box* survived in the shambles, containing a spirit, a voice to pronounce judgement. 'I can call spirits from the vasty deep,' said Glendower to Hotspur. 'But will they come when you do call for them?' asked the latter. They come for us today. They came pitilessly from the deepest depths of Watergate to cast down a President.

Just as we sometimes see things which are really visual hallucinations, so also we sometimes imagine that we hear things that do not exist outside our minds. Indeed, endless examples can be given of people who, suffering from, say, epilepsy, hear a great variety of voices and also of music which are entirely subjective. This is not to say that people who hear voices are necessarily ill. 'Criminal lunatic asylums are occupied largely by murderers who have obeyed orders,' writes Bernard Shaw. Their feelings or imaginings are so strong that the idea comes to them as an audible voice, sometimes uttered by a visible figure. 'Thus a woman may hear voices telling her that she must cut her husband's throat and strangle her child as they lie asleep; and she may feel obliged to do what she is told.' But you do not need to be criminally minded or weak minded or mad to see visions or hear voices. 'The inspirations and intuitions and unconsciously reasoned conclusions of genius sometimes assume similar illusions. Socrates, Luther, Swedenborg, Blake saw visions and heard voices just as Saint Joan did.' But I think it is doubtful whether Blake took his subjectivity very seriously since not only did he hear angels talking in a bush, but he saw the ghost of a flea. The voices which Joan of Arc heard were very practical and frightened her when she heard them first at the age of thirteen. The records of her six public examinations make it clear that one voice was so insistent that she felt she had no choice but to obey it. 'It told me that it was necessary for me to come into France. It said to me two or three times a week: "You must go into France." I could stay no longer. It said to me: "Go, raise the siege which is being made before the City of Orleans." ' The Voice was generally accompanied by a light. St Paul on the way to Damascus suddenly saw a great light and heard a voice crying: 'Saul, Saul, why persecutest thou me?' It is said that he was subject to 'the sacred disease of epilepsy' which is generally responsible for auditory illusion.

Of course the illusion of sound may come in many ways and be caused by means which have nothing to do with genius, criminality, or disease. There is nothing more charming than Kinglake's account in *Eothen* of how he heard the bells of his village church in Somerset while he was in the Sinai Desert. I came upon it many years ago, and it remained with me. I would like to share it. 'The sun, growing fiercer and fiercer, shone down more mightily than ever upon me he shone before, and as I drooped my head under his fire, and closed my eyes against the glare that surrounded me, I slowly fell asleep – for how many minutes or moments I cannot tell; but after a while I was gently awakened by a peal of church bells – my native bells – the innocent bells of Marlen, that never before sent forth their music beyond the Blaygon hills! My first idea naturally was that I still remained under the power of a dream. I roused myself and drew aside the silk that covered my eyes, and plunged my bare face into the light. Then at least I was well enough awakened; but still those old Marlen bells rang on, not ringing for joy, but properly, prosily, steadily, merrily ringing "for church". After a while the sound died away slowly. It happened that neither I nor any of my party had a watch to measure the exact time of its lasting, but it seemed to me that about ten minutes had passed before the bells ceased. I attributed the effect to the great heat of the sun, the perfect dryness of the clear air through which I moved, and the deep stillness of all around me. It seemed to me that these causes, by occasioning a great tension, a consequent susceptibility of the hearing organs, had rendered them liable to tingle under the passing touch of some mere memory that must have swept across my brain in a moment of sleep. Since my return to England it has been told me that like sounds have been heard at sea, and that the sailor, becalmed under a vertical sun in the midst of the wide ocean, has listened in trembling wonder to the chime of his own village bells.'

Indeed, why should they not? It seems most natural when we consider how strongly the mind can be affected by emotion and memory. 'O heavens bless my girl!' cries Pericles, Prince of Tyre, on at last discovering his daughter, Marina.

> But, hark, what music?
> Tell Helicanus, my Marina, tell him
> O'er, point by point, for yet he seems to doubt,
> How sure you are my daughter. But what music?

Helicanus – My lord, I hear none.
Pericles – None!
> The music of the spheres! List, my Marina.

Lysimachus – It is not good to cross him; give him way.
Pericles – Rarest sounds!
> Do you not hear?

Lysimachus – Music, my lord?
Pericles – I hear
> Most heavenly music!
> It nips me unto listening, and thick slumber
> Hangs upon mine eyes.

The fact that we see things and hear things which are not objective certainly does make it difficult to be sure about what actually *is* objective, if anything. The question which is generally posed runs as follows: if a tree falls to the ground with nobody in sight, does it exist, and if it exists, does it make any noise as it falls to the ground? I think that we are bound to accept that it does exist, even if by virtue of a kind of mad logic we could legitimately hold the view that it does not exist other than as internal sense datum. But when it falls all alone, does it make a noise? I suppose not. All that happens is that it makes vibrations in the air around. That is to say there has been molecular movement. But molecular movement is not sound: molecular movement is not light. The vibrational effects – or whatever we wish to call them – are interpreted by *us* as sound and light. Could we see the waves or vibrations, *per se*, we would not see redness or blueness, loudness or music. Yet when cells in our brains begin to be shaken in certain ways light and sound enter our minds. They enter our minds, or are construed by our minds, as sunlight and colour, as sound and music. If we could touch them or feel them, *per se*, they would not mean sunlight or song to us. The waves of ether have no colour, the molecules have no colour; the waves of the air are silent, the molecules make no sound.

Perhaps the problem was never better stated than by the great John Tyndall, one of the most readable of the Victorians. In a very famous address delivered at Belfast, he spoke as follows: 'I can follow the waves of sound until the tremors reach the water of the labyrinth and set the otoliths and Corti's fibres in motion; I can also visualize the waves of ether as they cross the eye and hit the retina. Nay more, I am able to pursue to the central organ the motion thus imparted at the periphery, and to see in idea the very molecules of the brain thrown into tremors. My insight is not baffled by these physical processes. What baffles and bewilders me is the notion that from these physical tremors things so utterly incongruous with them as sensation, thought, emotion, can be derived. You may say or think that this issue of consciousness from the clash of atoms is not more incongruous than the flash of light from the union of oxygen and hydrogen, but I beg to say that it is. For such incongruity as the flash is that which I now force upon your attention. The flash is an affair of consciousness, the objective counterpart of which is vibration. It is a flash only by your interpretation. *You* are the cause of the apparent incongruity, and *you* are the thing that puzzles me.'

That is what I meant when I remarked some pages back that 'luminosity is a very great mystery'. Meanwhile the debate continues. I will not continue it: not at the moment anyway. But I would like to round off this portion of my enquiry by returning to my point of departure when I talked about seeing, not with our eyes but by means of our eyes. For not only does perception make a philosopher's holiday, but optics makes a field day for the mechanists. A great physiologist once said to Ruskin that sight was 'altogether mechanical'. Now, in my opinion, Ruskin was more interesting and certainly more succinct on botany or geology or mineralogy or nature in general than as an art critic. He replied to the physiologist that his 'words meant, if they meant anything, that all his physiology had never taught him the difference between eyes and telescopes. Sight is an absolutely spiritual phenomenon; accurately and only, to be so defined: and the Let there be light, is as much, when you understand it, the ordering of intelligence as the ordering of vision.' Object as we may to the use of a word such as mind or spirit, we still have to account for luminosity. 'It is the appointment of change,' said Ruskin in *The Eagle's Nest*, 'of what had been else only a mechanical effluence from things unseen to things unseeing, from stars which did not shine to earth that could not receive; the change I say of that blind vibration into the glory of the sun and moon for human eyes.'

There is always the tendency to emphasize the mechanical at the expense of the spiritual. 'If now someone should ask you,' says Socrates to Theaetetus, 'by what does a man see black and white objects and by what does he hear shrill and low sounds, I suppose you would say, by eyes and ears.' Theaetetus assents and Socrates continues: 'A careless ease in the use of names and expressions without pedantic linguistic analysis is for the most part not ignoble. The opposite is rather the mark of a mind enslaved. But sometimes it is necessary to be more exact. For example, it is necessary now to consider in what respect the answer you have given is incorrect. Reflect: which answer is more correct, eyes are that *by* which we see, or that by means of which *we* see, and ears that *by* which we hear or that by means of which *we* hear?' Theaetetus answers, 'It seems to me, Socrates, truer to say, *by means of which* than *by which* we perceive in each case.' 'Yes indeed,' says Socrates, 'for it would surely be a terrible thing if so many powers of perception were seated in us like warriors in wooden horses, and if all these senses did not draw together into some one form, call it soul or what you will, by which, using these senses as instruments, we perceive whatever we do perceive.'

I still think it truer to say that we see by means of our eyes than that our eyes see; and if I have quoted from two nineteenth-century sages, and from Socrates, it is because I do not think that wisdom in this field is necessarily progressive, and I would contend that Socrates is just as up-to-date in this matter as the most agile of linguistic analysts sheltering under the shadow of Wittgenstein.

III Smell and taste

IN SPITE OF now being able to see and to hear, our man in the field is still very handicapped. It is not just that he can neither smell nor taste: he has no sense of touch, without which he could do nothing at all!

Let us think of smell first. I fear we take it so much for granted that only on reflection do we realize what we owe to it, and how underprivileged we would be without it. Imagine being deprived of the smell of wallflowers wafted to us in spring, or of stocks in summer. Try to think of roses without any scent. Consider the odour of soil after the plough has turned it. These smells carry with them the sense of earth's harmony and cadence while the message of decadence comes to us in the fumes of corruption. Doubtless this sense evolved as largely protective and functional, so that what is good for us and what is poisonous could be detected. Today we rely upon our stored-up knowledge and not our instinct to make this distinction. I go into a field and pluck a mushroom, and I know that it will be good to eat either on the spot or cooked. In the same field there may be a toadstool. I can tell by the pink interior of the mushroom that it is good, but the colour of the toadstool warns me that it is poisonous. I use the word poison lightly, but with a feeling of shame at my ignorance of the mystery. Why should 'my lord, the stomach' accept the one and reject the other? From one we get such pleasure while the other wrecks the gut. The animals, not having attained our intelligence, have retained their powerful instinctive sense of smell. Thus we hear of cats finding their way home over immense distances, while dogs can identify their master's tread even if inextricably mixed with the tread of other men. When Darwin's dog recognized him after an absence of five years and two days, and Ulysses' dog, called Argus, recognized him even through a beggar's clothing after an absence of twenty years, we may be very sure that in both cases the dog's recognition was not by eye but by nose. In the case of human beings the sense of smell as a function is fully brought into play only when the person is lacking in other senses. James Mitchell, born deaf and blind, could connect with the outside world by means of smell.* He readily observed the presence of a stranger in the room, and he determined his opinion of people from their characteristic odours of which others were

*See Dugald Stewart's *Works,* Vol. IV, p. 300.

unaware. According to Alexander von Humboldt some savage tribes have shown a sensibility no less acute: the Peruvian Indians were able to distinguish in the middle of the night whether an approaching stranger was a European, an American Indian, or a Negro.

We are aware of the existence of smell because not only do our noses carry out the breath-taking miracle of breathing, but they possess the olfactory nerve, the filaments of which bear a resemblance to a hand kitchen-brush. It is not in the mainspring of respiration but in a kind of cul-de-sac off the air passage, which explains why we must sniff (that is, breathe in quicker) if we wish to detect an odour. These filaments, or cilia, detect the smells, they are the net in which the scents are caught. Sherlock Holmes stipulated that a good detective should be able to recognize at least seventy-five odours out of the thousands that we meet with. He did not demand that the detective should know exactly how the cilia are able to do this, how they can send scent-impulses to the brain. For it has been found very difficult to nail down just what an odour *is*. Does the rose send us part of itself carried by the molecules of the air? But the authorities maintain that, for example, a bit of musk will continue to send out its scent for years perhaps and yet lose nothing in size or weight. Indeed the fineness of the particles is astonishing, because if the air conveying the odour is filtered through a tube packed with cotton-wool and inserted into the nose, a smell is still experienced! The How of this matter of smell has not yet, I gather, been completely explained. The Fact of it is scarcely appreciated enough. The sheer number of different odours, even from the same species such as the rose, is staggering. Is it the *essence* of life's children that thus declares itself? We are lifted up by the essence of lavender in the summer or of chrysanthemums in the autumn, while the essence rising from rotting leaves in the lonely wood and from the mossy mounds of fallen trees, last-flung shadow of the perished trunks, by no means casts us down.

It is interesting to note that one can experience olfactory

hallucinations no less than visual or auditory ones. People will become alarmed by smells which are clearly experienced, but nevertheless imaginary. Thus a patient of Dr Russell Brain experienced what he described as 'a smell of rubber burning' which would last for hours, quite objectively, he felt. He would wake up in the night and smell burning, convinced that it was an old beam in his farmhouse (such a beam can smoulder for days, as I have experienced myself). But at last he realized that 'it wasn't the house that was on fire – it was me!' He was not sure that it was 'Me' until he went somewhere else 'and found that I could still smell it'. A woman declared that she experienced her first attack of smell-hallucination while driving her car. She thought that the car battery must be leaking, so she stopped and opened the bonnet to look at it. It was in good order. She then realized that the smell emanated from within herself. And of course examples can be multiplied.

It is for the sake of methodology that we tend, perhaps too often, to separate the senses into compartments – but actually they intermingle a great deal. I note, in passing, that Lewis Mumford observed how the loss of coherent speech in senility 'even produced the illusion of blindness: what the eye beheld had become invisible: it no longer made sense.' Not surprising: for after all the person *is* a unity. Certainly taste is intermingled with vision and smell and touch. It is the palate which distinguishes the difference between strawberries and raspberries, but it must employ the offices of touch before doing so. Moreover, the colour of what we see often affects our sense of taste. Thus, as an experiment, according to Robert Fromen, a number of people were given mashed potatoes to eat in a dark room. They declared them to be delicious. By a harmless process the potatoes had been dyed purple. When the lights were turned on the sight of the potatoes instantly took away the appetite. Indeed our sense of smell is so linked with that of taste that if the olfactory nerves are prevented from functioning by simply holding your nose, taste will be so impaired that you cannot distinguish between an apple and a potato.

Since there is no planned educational drive towards developing the senses (a self-education that could continue throughout life) we probably see, hear, smell and taste about half of what is possible. We take less trouble to cultivate a sense of flavour than any other. It must be as galling to a person who has cooked a very savoury meal to see people gulping it down and talking their way through it as for an author to see someone reading his book, the while answering a child's question, or half-listening to and occasionally joining in the conversation of others around him. As it is, the cultivation of senses is almost entirely a question of need or of gifted inclination. The North American Indian could recognize footsteps and even in the dark distinguish between the tread of friends and foes, while his white companion was unable to do this. Nevertheless that same white man could have been a musician able to distinguish all the different instruments in an orchestra, which to the Indian would have been but a confused mass of sound. A person who has a great inclination to study wild flowers will be able to see ten times more than even a devoted country-lover, just as a bird specialist will be able to distinguish the note of every species of bird in the chorus which to others is but an assemblage of sounds.

We do not neglect our sense of touch because it is so important to us. Without it we could not put one brick upon another, lift a knife and fork, or walk across the street, let alone feel the signals of pain. When we bring the ends of our fingers into contact with an object an impression is immediately made, conveyed by the sensitive nerves to the spinal cord and thence to the brain. We say we have *felt* it – as sharp or smooth. This power which we possess of conceiving the nature of an object or a surface, without visual aid, is called the sense of touch. It is seated in the skin generally, but particularly in the fingers. It is not, however, possessed by the skin, but by certain structures in it which communicate with the terminations of the truly sensitive nerves. It resides in the extremities of these nerves. If a patient whose spinal cord has been severely damaged has the soles of his feet tickled, he will make no reaction. A healthy man would react at once, finding it extremely unpleasant. (Why should tickling make people *laugh*, especially children?) But with this man it is unfelt. We lay our hands upon other limbs and still he does not feel anything, and will declare that unless he had seen the position of our hands he would not have believed that they were in contact with him. The sense is absent because, the spinal cord being damaged, the line of communication between the surface of the limbs and the brain is severed, the telegraph wire has been cut, and the messages to the mind are no longer transmitted.

The importance, the desperate importance to us, however unconscious we are of it in the ordinary way, of the sense of touch, is brought home by hallucinations which sometimes arise when a limb has been amputated – a dreadful word. The commonest example of this is known as the 'phantom limb'. The person has the persistent feeling of the presence of the amputated limb. After losing his arm, Nelson experienced a phantom one and he regarded this as a proof of the existence of the soul. The person does not always actually see a limb which is not there, but the feeling can be so strong that a man with a phantom leg may fall down in an attempt to stand on it.

It used to be thought that bats had a *sixth* sense because of their capacity to avoid hitting an object in the dark, however swiftly they flew. W. H. Hudson held that it was a refinement or extension of the sense of touch – 'an excessive sensitiveness in the membrane'. He observed that blind men 'sometimes have a similar sensibility of the skin of the face. I have known one who was accustomed to spend some hours walking every day in Kensington Gardens, taking short cuts in any direction among the trees and never touching one, and no person seeing him moving so freely about would have imagined that he was totally blind.'

IV Touch

IT IS ONLY from the blind that we can really estimate the importance of touch. Indeed, unless we listen to what they have to say, our appreciation of the senses will be very inadequate. And when we think of the blind who have spoken to us on this subject we are bound to think of Helen Keller, who was one of the most gifted women of her time. Her book, *The World I Live In*, written for the most part (with a few florid lapses) with classic purity of style, is a revelation, and practically every sentence is enlightening. She was born blind and deaf, and, like all babies, unable to speak. For nearly seven years she lived in darkness and mental isolation. Before her teacher, another remarkable woman, came to her she did not know who or what she was. She lived in a world that was no-world. Her consciousness was only of nothingness. She had neither will nor intellect. She never contracted her forehead in the act of thinking. She loved nothing, she cared for no one. 'My inner life,' she declares, 'was a blank, without hope or anticipation, without wonder or joy or faith.'

What a wretched mind was there! An empty vessel indeed, adrift upon a shoreless sea; a thing appointed to desolation, for ever exiled from the world outside. But it was not so. Her deliverer came and she was redeemed from darkness; and the instrument of her redemption was the sense of touch. 'In my classification of the senses,' she was to write, 'smell is a little the ear's inferior, and touch is a great deal the eye's superior.'

She had been in prison, in a dungeon, until one day she found a little hole and gradually widened it until she could escape. Her powerful mind lay dormant until one day her teacher (her saviour) was able to make her equate the sensation of water with symbolic taps made into the palm of her hand. She grasped the existence of words. She learnt that a word stands for something, that each word symbolizes an object or a sensation or an event or a thought or an idea. In the end she was in possession of the world more fully than many people with five senses instead of three. I have said that she found a hole and escaped through it into the world outside her dungeon. Yet it is truer to put it the other way round: through that hole the world came in to her.

We learn so much from her about the senses: how, if a sense is maimed, sight for instance, then touch becomes more poignant and discriminating. 'It is not for me to say whether we see best with the hand or the eye,' she wrote. Our attention is held. She continues: 'I only know that the world I see with my fingers is alive, ruddy, and satisfying. Touch brings the blind many sweet certainties whith our more fortunate fellows miss, because their sense of touch is uncultivated. When they look at things they put their hands in their pockets. No doubt this is one reason why their knowledge is often so vague, inaccurate, and useless.' Severe words, but interesting. And notice how she uses the word *hand* – as she very rightly does since it is singular in its humanity. When writing a piece on 'The Seeing Hand' she opens with her dog giving her hand a lick. 'He pressed close to me as if he were fain to crowd himself into my hand. He loved me with his tail, with his paw, with his tongue. If he could speak, I believe he would say with me that paradise is attained with touch; for in touch is all love and intelligence.'

We are made to think twice. She knows that there are innumerable marvels unguessed by her; 'but there are a myriad sensations perceived by me of which you do not dream'. She declared that sometimes she felt the very substance of her flesh was so many eyes looking out at the world. The silence and darkness which were supposed to shut her out, 'open my door most hospitably to countless sensations that distract, inform, admonish, and amuse. With my three trusty guides, touch, smell, and taste I make many excursions into the borderland of experience which is in sight of the city of Light.'

We learn a great deal more from her about the strength of those other three senses, and of the extent to which all the senses assist and reinforce one another. 'The sense of smell' in her case 'became almost a new faculty to penetrate the tangle and vagueness of things'. She gives us an idea of the range of knowledge arising from odour, and she doubts if there is any sensation arising from sight more delightful than the odours from the earth. She found that human odours could aid her in recognizing people just as well as the recognition of hands and faces through touch. 'If many years should elapse before I saw an intimate friend again, I think I should recognise his odour instantly in the heart of Africa, as promptly as my brother that barks.' The range of her awareness would astonish those of us who think of the blind as unseeing and the deaf as unhearing. 'The thrilling energy of the all-encasing air is warm and rapturous,' she declared. 'Heat-waves and sound-waves play upon my face in infinite variety of combination until I am able to surmise what must be the myriad sounds that my senseless ears have not heard.' She held that in comparison with herself most people are smell-blind and vibration-deaf. 'Every atom of my body is a vibroscope,' she said. That was how she was able to *hear*. She could feel sound by means of vibration, and she draws up lists of sounds she could *feel*, including singing and music. It is startling to read her comment: 'The city is interesting; but the tactual silence of the country is always most welcome after the din of town and the irritating concussions of the train. How noiseless and undisturbing are the demolition, the repairs, and the alterations of nature!'

Her powerful mind could *unify* her impressions. That is what the human mind can do. A blind and deaf monkey could not do it. 'I tread the solid earth. I breathe the scented air. Out of these two experiences I form numberless associations and correspondences. I observe, I feel, I think, I imagine.' She could imagine: she could draw the exterior world into her mind. She could roam about in the landscapes of her interior world and perhaps improve upon the original. By virtue of her other senses she could use her imagination, *not* by giving to airy nothings a local habitation and a name, but in constructing her inner world on the solid foundation of what for her was purely sensational.

The great gift bestowed upon the blind by Louis Braille made it possible for such a person to become very widely read. In one 'show-off' essay she reveals herself as a sophisticated, formidable, and amusing bluestocking. She has such a taking way with her that we are charmed into accepting whole-heartedly her strictures upon the poverty of so many people's inner lives, people who, seeing nothing, think there is nothing to see; and we respect her pity for those who are less privileged than herself in this way.

Finally, two more striking remarks she made about our senses. 'The senses assist and reinforce each other to such an extent that I am not sure whether touch or smell tell me most about the world.' And: 'I am sure that if a fairy bid me choose between the sense of light and that of touch, I would not part with the warm, endearing contact of human hands or the wealth of form, the nobility and fullness that press into my palms.'

If I have used Helen Keller as the best eye-opener on blindness and ear-opener on deafness, and for instruction about the senses in general, it is not because she stands alone but on account of her gift of expression. The following verse, called 'The Mountain and the Pine', was written by Clarence Hawkes, who was blind from childhood.

> Thou tall, majestic monarch of the wood,
> > That standest where no wild vines dare to creep,
> Men call thee old, and say that thou hast stood
> > A century upon my rugged steep;
> Yet unto me thy life is but a day,
> > When I recall the things that I have seen, –
> The forest monarchs that have passed away
> > Upon the spot where first I saw thy green;
> For I am older than the age of man,
> > Or all the living things that crawl or creep,
> > Or birds of air, or creatures of the deep;
> I was the first dim outline of God's plan:

> Only the waters of the restless sea
> And the infinite stars in heaven are old to me.

Such can be a blind man's power of vision.

If I have left Milton out of these considerations it is because he went blind at the age of forty-three, and was in no wise saved by the sense of touch. Who can forget his words in *Samson Agonistes*: 'Dark, dark, dark amidst the blaze of noon,/Irrevocably dark, total eclipse/Without all hope of day!' Perhaps we may pause for a moment to listen to the words, in *Paradise Lost*, that sprang from that disaster, and thus remind ourselves of his enormous stature.

> Hail, holy Light! offspring of heaven first born,
> Or of the Eternal co-eternal beam,
> May I express thee unblamed? since God is light,
> And never but in unapproached light
> Dwelt from eternity; dwelt then in thee,
> Bright effluence of bright essence increate.
>   ... Thus with the year
> Seasons return, but not to me returns
> Day, or the sweet approach of even or morn,
> Or sight of vernal bloom, or summer's rose,
> Or flocks, or herds, or human face divine;
> But cloud instead, and ever-during dark,
> Surrounds me, from the cheerful ways of men
> Cut off, and for the book of knowledge fair
> Presented with a universal blank
> Of nature's works, to me expunged and rased,
> And wisdom at one entrance quite shut out.
> So much the rather thou, celestial Light,
> Shine inward, and the mind through all her powers
> Irradiate; there plant eyes, all mist from thence
> Purge and disperse, that I may see and tell
> Of things invisible to mortal sight.

We Communicate

I Speech

IT MIGHT SEEM that at this point our task is over. Our man in the field has got everything necessary in order to carry on. Yes indeed, he can carry on now – but only as an animal. Everything we have given him so far has also been given to the animals. As it stands he in no way differs from them in basic acquirements. But we know that he does differ enormously. He possesses another faculty. All animals have mouths, and he also has a mouth in order, like them, to put into it a portion of the earth so that he may remain part of the earth. But he has employed his mouth and tongue and larynx and lips for a function inaccessible to the animals. So we come to *speech*. Not just the capacity to bark or growl or screech, but to acquire *language*.

It is very difficult to think freshly of the first men. Sometimes capitals are used and we speak of Early Man. That doesn't help much; but Early Man was Early Man because he began to speak. To us now it seems a natural advance. We can understand a man making a noise to another man in the distance, a noise sounding like 'Hi!' And we can imagine the sound 'Hi' becoming eventually more like 'Here', and then after a lapse of time becoming 'Here I am!' It does not seem a very surprising development. What is so surprising is that none of the other animals followed a similar line. Why could not the gorilla, say, or the chimpanzee, or the orang-outang, or the gibbon, make an advance in this manner? They have not done so. To this day they are stuck with their primary sounds. 'What sadness is in the eyes of monkeys,' said Bertrand Russell, 'as if they felt they had lost their way in evolution.'

Is it just a question of physiology, an insufficiency of what we call 'grey matter' in the skull? But we are told that the quantity of grey matter is not the chief thing, and we have examples of idiots sometimes possessing more of it than exceptionally brainy people. The matter in the human cortex is built up with five distinct layers of cells: it is upon the complexity of their arrangement that the human intellect depends. I quote from Jacquetta Hawkes who, after suggesting that perhaps the amount of grey matter may to some extent be the answer, adds: 'When a cat looks at a king he can draw upon no more than three layers for his conclusions, and this limitation, although undoubtedly it leaves him more content in his untroubled bodily perfection, will prevent him from ever being able to publish a book, or even give a lecture, on the subject of royalty.'

How many of us have wished that we could communicate with a dog to the extent of being understood if we said 'Back on Monday' or 'Back two days' or just 'Back' – or the gesture of two fingers held up as indicative of two days. But it is hopeless. The animals are unable to receive a conception. They are conscious without consciousness. They cannot conceive time as such, nor space as such. There is a barrier they are unable to cross and though centuries roll by they never do cross it, they never speak for they cannot conceive.

This barrier irks us; we long to overcome it, for we know that we share our ancestry with the animals. We like to pretend that they can speak. Hence the enormous popularity of *The Wind in the Willows*, and of Swift's horses and Orwell's pigs. No book has been more loved in modern times than Joy Adamson's *Born Free*. For that remarkable story did disclose a sense of rapport between man and animal which appealed to everyone. There is one striking vignette, when Elsa, the lioness who has been brought up with the Adamsons and then gone into the wild, comes to see them whenever they make a passing visit in her vicinity. 'For some time before the final parting she becomes noticeably aloof and turns her face

124

away from us; although she wants desperately to be with us, yet, when she realises that we are going, she makes it easier in this touching, dignified and controlled way. As this happens every time, it can hardly be coincidence.'

It was almost as if the barrier of communication had been removed, as indeed it had; but the main barrier remains – we cannot converse with them. And they do not converse with one another. They do not wish to do so, they have no need to do it. See those two horses in the field, such good friends, eating grass side by side, necks close together, nose by nose. They communicate in their own manner, but without conception which needs language as an instrument. The jockey who won the Grand National this year spoke warmly of his horse whom he represented as being well-read with regard to the course, doing most of the head-work, and being delighted with the applause of the spectators. Now, we know that animals can and do frolic together and run about with sheer *joie de vivre*, but they have no mental conception of 'play' or 'game' or 'race'. It is not to be supposed that a single horse at the Derby has any idea of a race as such or knows that he has 'won' or 'lost'. Horses apparently will never develop the capacity to enlighten us about this, just as a lot of monkeys jabbering like a lot of monkeys will never jabber themselves into some conceptual remark.

A baby at birth is less articulate than a monkey. It knows nothing of speech just as it knows nothing of space or of time. Yet by the age of four the child is quite at ease with language structure without laboriously learning it. 'In slightly more than two years children acquire full knowledge of the grammatical system of their native tongue,' surprisingly comments Professor David McNeil of Michigan. They are at home in a world of sounds of every variety, and are able to imitate them. 'At birth,' writes Anthony Smith in a striking figure, 'a child not only knows no speech but, unlike an adult arriving at Peking or Ulan-Bator without a phrase book, has no concept of speech, no preconceived notions whatsoever, let alone such refinements of grammar such as "bite dog" being so different to "dog bite". It is a stunning intellectual achievement, unthinkable by most adults, and yet carried out by every child even before it reaches school.' That is the uniqueness of mankind and sets us apart from all the animals who are unable to command more than variations of a grunt or a growl or a neigh or a roar or a howl or a purr, while 'birdsong' does not actually reach the condition of singing. Think of the range of the human voice in timbre whether in song or speech. The whole character also is there. I have often heard it said that people can best be judged by their voice.

The physiological basis is more elaborate than the sound apparatus of the animals, and lends itself to a wide degree of conscious improvement. The main properties can be simply stated: we have the windpipe (or trachea) which possesses two branches (or bronchi), at the upper end of which is a structure of muscle and cartilage called the larynx containing the vocal chords that are made to vibrate when air from the lungs is forced through them. They are wind-instruments. Actually chords is a less accurate term than reeds would be; and seeing what they do for us, reeds is certainly the most appropriate and evocative term, carrying us back in history to John the Baptist who declared: 'I am a voice crying in the wilderness,' and of whom Jesus said: 'He is as a reed shaken by the wind.'

The pitch is controlled by varying the tension on these instruments, and by regulating the air passing through the larynx. No Prime Minister can make a real impact without a good one. It was an essential instrument with which Gladstone, Lloyd George and Winston Churchill could make history. It didn't matter what Gladstone said: his audience was riveted by his larynx.

I have suggested that our apparatus lends itself to conscious development. Indeed enormously so – and we enormously neglect it. Most of us out of laziness deploy perhaps only a third of our potential in tone and articulateness. A pig does not

126

know how to operate its mouth so as to say 'P' for pig, nor a
dog how to operate its tongue so as to say 'D' for dog. We know
these things but we seldom make the best of them. And
making the best of speech also involves proper breathing and
deep breathing which is so important to health. Successful
actors and actresses live into their eighties or nineties because
they tend to take breathing exercises every day. It is necessary
for their job. It is their breath of life. It is box office.

At any rate the voice is the most magical of our attributes.
There is magic in Hilaire Belloc's appeal:

> Tonight in million-voicèd London I
> Was lonely as the million-pointed sky
> Until I heard your voice. Ah! so the sun
> Peoples all heaven, although he be but one.

II   The evolutionary process

How IS IT that mankind alone possesses the power of speech?
The short answer is that we alone possess intellectual
consciousness, and have invented words to give expression to
the faculty. But a longer answer seems obligatory.

In the beginning is the end, as in the end is the beginning. I
use that phrase as a bid to avoid using the word evolution
which has become so worn that it scarcely makes an impact.
All the same, I must resolutely use it if we are to see our man in
the field clearly and account for the birth of language. 'I lead no
man to a dinner-table, library, exchange,' said Walt Whitman,
'but each man and each woman of you, I lead upon a Knoll.' Or,
as the historian, and endearing Oxford don, Dr Ernest Barker,
used to say: 'Perspective, gentlemen please . . . perspective.'

Let us start with the acknowledgement that the evolutionary process has been conceded. The fact that you are probably
surprised that I should come up with so banal a statement is a
measure of the change in the world-picture that has taken
place in only a hundred years. For people did actually believe in

the Garden of Eden, not just till the Middle Ages but well into the last century, and they did see man as completely separated in ancestry from the animals. But even now we have not yet quite settled down, as it seems to me, to an agreed interpretation of the significance of the process. When Samuel Butler said in his *God the Known and God the Unknown* (a brief adjunct to his *Evolution Old and New*), 'We must therefore see the whole varied congeries of living things as a single very ancient Being, of inconceivable vastness, and animated by one Spirit,' he was thought to have made a complete statement for modern minds. But it was not complete. We are obliged to include the primeval soup. We must go further back and include the first gases and rocks as well as 'living things', and drive no wedge between them.

Thus our cosmic picture is: first fire, water, and gases; then rocks; then plants; then animals; then man. In short, a unified unfolding seemingly progressive and purposeful, though we have no idea of how it could ever have begun. It could have started as the invisible taking on visibility, or pure thought which then declared itself as matter (after all, the thought of a house, say, the idea of a house, must always come before the actual existence of the visible house). However, since speculation on the first cause has been found to be unrewarding, thinkers are content to deal with the finished article – the manifest process and its evolutionary appetite – and to mark certain stages in that process; the stage or jump from the inorganic to the plant which grows; from the plant which grows to the animal which moves about; from the animal to the man who becomes conscious of the whole thing. Thus we get a single process up to the emergence of mind. We do not see man as in the earth, or on the earth, but as a piece of the earth.

It is possible to regard this matter in what is called a mechanistic light. You can say that everything is based entirely upon mechanistic laws, that there is no purpose in the affair, no guiding hand, that it is simply a question of accidental variation in a process of natural selection in which the fittest survive, and that the general machinery includes a strong reproductive impulse and a great tendency to variation. You can say that the material world is alone basic and permanent, and that life, mind, and purpose are temporary phases which have accidentally occurred at a certain point in time and may disappear at some future time through the mindless and purposeless operation of the mechanical laws of the basic material world. Indeed there are some people who are so enamoured of mechanism that they are ready to dismiss certain human beings as having *a screw loose*.

I am a great admirer of accident. Take that organ of the human physiology which we have already studied – the eye. Its retinal surface, which is sensitive to light, can distinguish between vibrations of 450 million million a second, which give the sensation of red, and 750 million million a second, which give the sensation of violet. It can pick and choose in this manner with the greatest of ease and can do many other things just as remarkable. If it does this by accident or 'accidental variation', I must welcome accident and subscribe myself his humble and obedient servant, and would wish that he had a more impressive name such as Mighty Spirit, or God, or Miracle-worker.

When terms are used such as 'survival of the fittest' or 'natural selection' or 'sexual selection' or 'reproductive impulse' or 'tendency to variation', we find on examining them closely that they are often merely lame descriptions and not an explanation. I say lame description for the survival of the fittest simply means that those survive who do survive, while to say that the peacock spreads such prodigy of plumage in order to attract the hen is plainly inadequate, since two bright feathers would be enough.

Nearly all the significant words used to support a mechanistic view of the world 'were originally merely tentative terms used by necessity to *describe* processes that were as yet inexplicable,' writes Professor Wilson in his *The Miraculous Birth of Language.** 'But these tentative terms by repetition and consequent familiarity have tended more and more to become accepted as *explanations* of the phenomena rather than tentative descriptions. Hence their tendency to conceal without explaining the problem.'

It is valuable to realize that description is not explanation. To get this distinction really into one's head requires some effort. In my twenties and thirties I spent a long time 'searching for Truth', rather frantically. (I did not find it, of course, I found something else.) In the course of the search I did at least come to realize, and to find some freedom in the realization, that just as description is not explanation so also information is not knowledge. In a book called *Farewell to Argument* I concluded a paragraph with the words: 'Information, however passionately pursued, however cherished by those who have fought for it through sweat and pain and hardship, however delicately built up, never becomes Knowledge, never becomes Truth, but remains information to the end of time.' Though crudely over-emphasized, that still seems to me sound, and I am glad to return to it now that I am dealing with the body, when description of its parts does not explain its unifying principle and purposeful design.

III   The coming of the word

WHAT IS THE VIEW taken by those who reject the mechanistic approach? They accept the mechanistic laws while regarding them as the means by which the vital force, or primal energy, expresses its potential. Matter is seen not as an independent substance but as the vehicle by which life and mind work out from the potential to the actual destiny. This inherent energy, or mind-force, is there at the beginning, and gradually, no

---

*See Bibliography.

matter how many aeons pass, moulds the material into what we call inorganic formations, and then after more centuries into organic forms, and finally into the form of man who, standing erect and ready to use his front legs as arms and his front paws as hands, develops a mind which can use those arms and hands to further ends, and at the same time consciously view the whole process! 'It is a long way from granite to the oyster, and farther yet to Plato,' said Emerson. 'Yet all must come as surely as the atom has two sides.'

According to this view, so far from mechanical law being at the basis of things, life and mind is the basis giving significance to the mechanics, and what is called inorganic matter. 'Life, mind, and purpose, latent and potential in the pre-organic period, emerge to actuality in the organic world, in an ascending series from plant to man.'

This is also only an hypothesis, not a proven explanation. But at least it makes sense, whereas the 'accidental' hypothesis does not. Clever men who are capable of making out a case for purposelessness will always do so, for it is good fun. But it seems perverse. 'Many a scientist has patiently designed experiments for the *purpose* of substantiating his belief that animal operations are motivated by no purpose,' wrote the celebrated A. N. Whitehead. 'He has perhaps spent his spare time in writing articles to prove that human beings are as other animals so that purpose is a category irrelevant for the explanation of their bodily activities, his own included. Scientists animated by the purpose of proving that they are purposeless constitute an interesting subject for study.'

It may be maintained that to talk about creative evolution nowadays is outmoded. That may be so. But I am dubious about the importance of modes. What is true on Monday is likely to be true on Tuesday, and what is outmoded on Wednesday can still be sound. In order to support mechanists and behaviourists who refer everything 'merely' to mechanical and chemical laws you must disregard the views of Plato, Herder, Goethe, Kant, Darwin himself, Tyndall, Carlyle, Emerson, Bergson,

Whitehead, Julian Huxley, Samuel Butler, Bernard Shaw, Teilhard de Chardin, to pick but fourteen names. The name Goethe is significant here, for he was less a poet who was interested in science than a scientist who felt compelled to express his scientific philosophy in poetical terms. Several large rooms in Goethe's house at Weimar contain nothing but the apparatus and materials for scientific experiment. Chemicals, ores, intricate instruments occupy table after table, while his library contains many more books of chemistry, physics and geology than of literature. But he expressed himself in poetic terms for he knew that it is unscientific to be only scientific just as it is unreligious to be only religious. It is not sound for the scientist to ignore the data arising from the experience of thousands of people, any more than it is sound for a religious person to ignore scientific data.

And if I have reintroduced the term creative evolution, I have not used the adjective loosely or weakly. I have been careful to include the primeval soup. In his preface to Professor Wilson's book *The Miraculous Birth of Language*, Bernard Shaw drops the remark that it was clear enough to him in his teens that 'if a dissolved salt can crystallize itself into a solid stone it is as much alive as the nearest squalling baby.'

Bernard Shaw, who never got the drift of Wordsworth, would not have enlisted him to support his evolutionary ideas; but it is interesting to note how often during the last forty years Wordsworth has been cited by scientists and scholars in support of their views. For Wordsworth's attitude towards Nature was not natural. The ordinary response is joy at the beauty, pain at the cruelty, and fear of lonely places. But Wordsworth went back to the primeval soup and nothing stirred him so much as rocks. He did not respond only to the organic; he felt that the inorganic was part of the pyramidal world structure of which man is the apex. It is only at the apex that mind emerges; but to Wordsworth it is already present at the base, and could be *felt*. He did not consider that the sky or the air or the ocean were mindless phenomena, but that they were actually animated by mind which we can feel if we are sufficiently sensitive. *Tintern Abbey* has been so over-quoted that it is difficult to read it straight, as it were, and see what it is actually saying. *The Prelude* has fared better, and Professor Wilson's remark can be read with an element of surprise: 'The *Prelude*, as Coleridge pointed out, is fundamentally a philosophical poem, and is in a very real sense the original English text upon which all the expository work on creative or emergent evolution of the present century is based, however unconscious the exponents may have been of Wordsworth's pioneer work in this field.' The good of saying a thing in poetry rather than in prose is that the words convince because there is a spirit abroad. If someone says that matter and mind make a vast unity with an evolutionary appetite that will always prevail in spite of calamity, we may agree or disagree in an academic mood, but we are beguiled into positive surmise when we read, as from *The Prelude*, Book V:

> Should the whole frame of the earth by inward throes
> Be wrenched, or fire come down from far to scorch
> Her pleasant habitations, and dry up
> Old Ocean in his bed, left singed and bare
> Yet should the living Presence still subsist
> Victorious, and composure would ensue,
> And kindlings like the morning – presage sure
> Of day returning and of life revived.

Take a similar statement from a contemporary poet – Ruth Pitter. However strangely, and obliquely, the pen is slanted, we get the meaning, for we feel that the author is authorized by the very spirit which is addressed. The poem is called *The Primordial Cell*.

> Think, muse on her, do not forget our common mother.

> She is to me as the old withered mother of many
> Spinning beside the doorway of her crumbling cottage,
> Dreaming of those long fled from her, those nobler faces,

And glorying darkly in her children's fairer children:
Too humble to look for gratitude; and O too sunken
In aged-infant dream to care that many suffer:
Poor beyond reach of pity, and in huge indifference
Wealthy, though all her progeny die in the world's ruin.

Glorious, and justified eternally, is our mother.

For if we perished all, she would arise a virgin,
Immaculate as on the third day of creation,
Capable of bearing new gods and greater heroes,
Conceiving a more ravishing rose, a lovelier lily,
Determined to outdo our vaunted glories, she,
Triumphing over our vacant places, would nobly fill them.

She eats disaster: war and famine are her plough-oxen.
Ruin this star, she would partake her to another.
Down dread eternities she looks, not for perfection.

She spins forgotten at the doorway of her cottage.
Think, muse on her, do not forget our common mother.

Do these considerations fall outside our present physiological inquiry? I think not; for we are swiftly approaching our man in the field with the faculty which separates him from any other creature on the earth – the sound that comes from his lips in the form of speech.

I have been assuming, as the hypothesis which makes sense, that the life-principle (not subject to investigation or explanation itself) presses forward to ever more complicated and higher manifestations of itself, however much, and however often, this form and that form may be due to environmental obstruction or selected by natural pressure. I use the word 'higher' as it seems justifiable: the rock may not seem higher than the gas arising from the raging riot of the water and the fire, but we do think of the plant as higher, and the animal as higher still, and then man as a step beyond.

After the inorganic, in the ascending scale, we see that plants have attained something new. A stone does not possess the power of growth. It cannot eat the earth and increase its size. The plant can partake of the earth and of the sky; however stony seems the ancient acorn it can raise itself above the ground as no rock can ever do, however latent lies the primal principle. The plant has attained that amount of individuality, that amount of freedom. The animals attained something more; they could not only partake of their environment but also move about in it; another step in individuality and another step in freedom from absolute mechanical laws when objects are moved only by external impact. They could do more than that. Unlike the plants they became aware of the world outside themselves, they attained senses through which they received intelligence that there was space around them, and they could see and hear what was necessary for their survival. And they could do more than that: they could make *sounds*. One animal could communicate with another animal. It could make a cry or a call or a yelp of fear. These acquirements gradually emerged during thousands of years of evolutionary appetite. Then came an emergency which was so acute that the animal ceased to be an animal and became something else! It had all the trappings of an animal, but intelligence had developed into mind. So we come to Man.

It is interesting that we can call him man, since before this nothing in all the world was called anything by anybody. To call a thing something was the answer to this new emergency – the state of consciousness. This was another step in individuality, another step in freedom. It was a greater advance than had ever been made before. What happened was that this new species, man, became aware of space as such, and saw objects in space so clearly and objectively that he was obliged to differentiate one from another *in his mind*. We call this faculty – by virtue of which we say 'that thing over there is a tree' – the capacity for conceptual thought. This capacity is ours alone, and has served to separate us completely from all other creatures. As to why this faculty emerged in one animal and not in another, this is still a matter for debate. Dr Julian

Huxley, in his book *The Uniqueness of Man*, contends that *only* the physiology of the primates (lemurs, apes, monkeys, gibbons, gorillas) with their upright posture could promote the evolution of conceptual thought, and that only in man did the erect posture become so absolute that thought could prevail.

It seems to me that we are obliged to see the matter in an evolutionary light and to regard man as the vessel long prepared to take on eventually the burden of consciousness so that Nature herself could look at Nature. In this matter I profess myself a confirmed Kantian. Of course, since he wrote the coping-stone to all his work, *The Critique of Judgement*, many other mighty names have said the same thing, just as others had done so before him, such as Plato with the main idea. 'Shall we not maintain then,' he wrote in *The Laws*, 360 BC, 'that Mind is the first origin and moving power of all this, or has been, or will be, since it has been clearly shown that Mind is the source of change and motion in all things?' Kant's view was that everything in the world has been produced by a purposive mind-force shaping all sorts of diversified forms to work out a potential destiny, that there is a total inner purposiveness working its way in matter or through matter from the lowest forms to the highest form in man. When man is reached, then this unconscious purposiveness becomes conscious and begins to carry on the work of evolution consciously with understanding of itself. Man is the vessel prepared throughout all the foregoing aeons for this moment, he is the apex and point of the whole thing. 'Plants are evidently for the sake of animals and animals for the sake of man; thus Nature, who does nothing in vain, has done all things for the sake of man, who is the crowning end, purpose, and final cause towards which all has been tending.' That is not from Kant but from Aristotle, for the intuition regarding purpose belongs to thinkers of all ages. If at the same time there are always severe semantic philosophers who juggle with words to prove that we are juggling with words, at least that serves to keep us on our toes and to warn us from claiming more than speculation, while we rest in the assurance that there is a certain soundness in preferring sense to senselessness.

We see three momentous phases or turning points in the process of this evolution. There was the 'moment', no doubt taking centuries, when, speaking very figuratively, something stirred upon the stone, and began to grow: a bit of the earth standing up and falling down, growing and withering away and growing again. Then the moment when something stirred upon this living thing, this plant, and detached itself and moved about; a bit of the earth moving upon the earth, with a nervous system and a headquarters which co-ordinated its movements. Then the moment, the most momentous of all, when something stirred within this moving creature, and its rudimentary intelligence became a mind, a faculty that could conceive the actualities of time and space instead of being imprisoned by them. This new faculty, consciousness, needed an instrument to make good its potentialities – and this instrument was the *word*. The ultimate force, the shaping principle, had worked in an embryonic fashion, moulding material until it was possible for the word to be used; but it was there at the beginning; and if we do take this emergent view of life then we can truly say, in biblical phrase: In the beginning was the Word, the Word was with God, nay the Word was God.

IV The naming of things

I HAVE JUST defined mind as 'the faculty that could conceive the actuality of time and space instead of being simply imprisoned by them.' Until the appearance of the first man no creature had any conception of space or time because they had no concepts of any kind. As for space, the eagle with its eagle-eye could see its prey at a certain point in space, or the lion pounce upon the running wildebeest, but they had no conception of space as such, and of the innumerable *different* objects therein. Your dog is as indifferent to the difference between weeds and your best flower-bed as he is incapable of distinguishing in your study the pen, the books, the table, the chairs, the clock, the sofa. This last, the sofa, he may discern for he may associate it with comfort just as the ball you throw for him on the beach is discerned as something to run after. A horse is sensitive to its immediate surroundings, and can be alarmed by the sudden appearance of an object such as a car, but has no conception of 'car'. On every road in the Highlands of Scotland sheep will wander across indifferent to the cars, having no conception of them, however many may pass (this causes no indignation in the drivers, as such human behaviour would do, for their lack of conception pleases rather than annoys us). Or think of the matter in terms of spaciousness, of distance. Distance is an idea, and animals do not have ideas: a horse that has won the Derby in the South of England could not conceive of some other horse winning a race somewhere else. Time is very much an idea, and they can have no sense of it; it descends upon them as cold or heat does. They are enclosed by it; they are enclosed by time and by space.

The first man took a great leap and conceived time and space and began to differentiate one object from another in an intellectual way. The fact that this leap was very far from being a swift jump, and may have taken many thousands of years of evolutionary endeavour to accomplish, is irrelevant to the fact that a leap was taken. And having taken it what was the first thing he did? Having conceived that there was such a thing as a definite object, many definite objects, what did he do? He fastened them down by *giving them a name*. The first word must have been a noun; and so important a position does it hold in our vocabulary that the re-nouncing of something or somebody carries with it imponderable reverberations. Nouns are stakes to which objects are tied, definitely fixed so that the objective world may be defined. Thus today we will not find anything without a name, or if something is found without a name it is regarded as a great discovery. And early man not only separated and named objects but he gave names to his fellows, and to this day it is impossible to find anyone without a name: *What is his name?* we ask as if it were the most natural thing in the world instead of something quite new in the calendar of nature – imagine each ant in a formicary or monkey in a monkery possessing a name. 'Adam's first task was giving names to natural Appearances: what is ours still but a continuation of the same?' asked Carlyle. And since one of the main characteristics of mankind is going to extremes, we have taken such delight in putting things into classes, in classification, that subjects such as botany and physiology are so overburdened with nomenclature it is often hard to see the wood for the trees.

This momentous event in evolution when one member of the species began to name things was first given literary expression by the Hebrew writer of *Genesis*: 'And out of the ground the Lord God formed every beast of the field, and every fowl of the air; and brought them unto Adam to see what he would call them; and whatsoever Adam called every living creature, that was the name thereof. And Adam gave names to all cattle, and to the fowl of the air, and to every beast of the field.' When we are young and sophisticated we are inclined to reject this as a silly way of putting it but when we are older we are more ready to accept a fable as the fairest way to formulate a profound truth. I do not care for any part of Bernard Shaw's *Back to Methuselah* except the first bit called 'In the Beginning', which is on a higher plane than the rest of the work. It is told in terms of the fable of the Garden of Eden, and the Serpent keeps finding *words* for each new idea that occurs to Adam and Eve. It is as profound as it is simple in the telling, and as dramatic as it is beautiful.

At the same time we must remember that while a fable may be the best way to convey the significance of a momentous event, nevertheless such simplification may obscure the

139

labour involved in bringing that event to pass. The birth of language was a slow, tortuous, and terribly laborious affair taking many thousands of years to accomplish. A name, I have suggested, is a stake to which we tie an observed object and thus keep it fixed in the flux. But at first *one* name was by no means always given to one object. At one time, according to Herder, the Arab language had a great many synonyms for familiar things: 'fifty different names for the lion, two hundred for the snake, eighty for honey, seventy for a stone.' In his book *The Myth of the Machine* Lewis Mumford has made a determined effort to give some idea of the labour involved in the creation of language and to expose the absurdity of regarding the advance of early man in terms of 'tool-making'. His tools remained wretched for thousands of years, for nearly all his effort was put into word-making, and only when that had been accomplished could he really make tools and subsequently civilizations. If archaeologists had been able to discover deposits of *words* as well as artefacts, how much fuller a story there would be to tell!

V   The birth of language

WE ARE NOW at liberty to return to our man in the field, and by way of summary, together with greater detail, tell the story of the growth of his special gift of language – itself a tale as thrilling as any which he consequently could narrate.

Let us see just what happened when the last animal, as it were, became the first man. It was then that he was loosed from the bonds that tied him down and he entered a new kingdom. He remained part of the earth as he had always been and lived in it as he had always lived; but now he added to this a mental conception of it which gave him another dimension of individuality and another lease of freedom. He looked round and saw things as they had never been seen before, as if scales had fallen from his eyes and chains had been struck from his feet. He looked round and saw what things constituted space. He saw animals, forests, rivers, mountains, oceans, clouds. He formed a definite conception of them as separate objects. He was now in possession of *two worlds*. There was the actual world around him composed of all these different things, and there was his mental conception of them. There was the earth itself and the image of it which he had in his head and which he could carry around with him wherever he went, and which would remain with him even if the counterpart had changed or vanished. He had never before possessed the power to make a mental image, never known an inner world that was the facsimile of the world outside now apprehended for the first time. And while he began to realize the meaning of space and to explore the kingdom that opened before him, he also achieved a conception of time. He began to nail it down, to watch it, to get the measure of it. That is to say he was no longer the slave of time but could bestow order upon it by marking it. To this day every sergeant in the Army knows that the best way to impose discipline upon his platoon is to issue the irksome command to *mark time*, just as *doing time* in prison is the most terrible way of wasting it.

Man had escaped from the bondage of space and of time. He had been born in chains but now had achieved freedom; he had stormed his Bastille and the two great walls of his prison had fallen to the ground. But as yet his kingdom was empty. How could he celebrate his delivery and prepare a path for the potential? How could he communicate with his fellows and take joint possession of the dominion and share in its trophies and its spoils?

Perhaps gestures would do? They are useful, of course. It is useful even today if you meet a gorilla in the jungle to know that nodding the head signifies peaceful intentions to the mind of the gorilla, but gesture would not cover enough ground, embrace enough objects, describe events, or ventilate ideas – which would be the property of the kingdom which he had inherited. Moreover, gesture is no good in the dark, nor from behind. How about the pictorial, a picture? Very good indeed at imparting a certain kind of information, and casting a certain kind of spell, and making an exactitude of communication in terms of magic or religion or just hunting. The cave-drawings are proof of early man's extraordinary expertise in this mode (though we must not think of 'early man' in the round doing such things, but rather of a few very gifted men in that age, as in all ages, creating in the midst of savages and philistines). But neither painting nor sculpture was sufficiently *portable* to serve well as a medium for communication. How about ritual and dance? Yes, much communication was made and is still made in this mode; but neither dance nor ritual could serve the mandates of the mind craving its dominion. How about music? We know the scope of its realm and the magnitude of its power to fathom the wordless depths, to stir the emotions, and to lighten the heart. The prolonged applause accorded to musical performances, greater than applause ever granted to any other medium, is as significant as it is marvellous to hear. But as an instrument of intellectual communication music is not prosaic enough, and far too indefinite in its articulation.

Where then was he to turn? What instrument could he use to furnish the kingdom he had entered? It must be done by means of his *voice*. He must learn to utter certain sounds. They must be sounds of a particular nature. In short, a new kind of sound. Not an inarticulate cry or groan or hail but something definite and conventionalized. Not Woo! or Whoop! or Hi! but DOG or BIRD or CAVE or FOOD: sounds which referred to one definite thing and not to any other thing. And this is what he did, he made the shapes in space definite by defining them with arbitrary words. A significant sound is passed from the lips of one man to the ears of another man, an invisible vibration, but with enough power in it to promote vast intercommunication between peoples. This shaped sound appears, and already civilization has been decided upon; and with the aid of the *fingers* the mind is ready to introduce all the cultures in the history of mankind.

Even so, though a good beginning, oral sound alone was not good enough. It could never do justice to the demands of mind. The life of the spoken word could only continue according to the memory of those who heard or uttered it. It was as perishable as the flying vapour or the fleeting cloud. At least the drawing in the cave, the imperial poise of the stone statue on the lonely island, would carry their message through successive generations, but the verbal message, the verbal description or idea, is no sooner uttered than it passes away. What could man do to make speech permanent and not confined to the vanishing vibrations of sound? He could write it down.

Write? What's that? We are going too fast. He could make a sign. He would not dispense with sound, but he could do without it if necessary – and do better. He could take a sharp instrument with his fingers and on a piece of material he could make a sign that would stand for HORSE or DOG or MEAT. But how make a sign that would signify an object? It would have to be pictorial to some extent. And indeed that is how the written word started. A snake makes a hissing noise, SSSSSSSSSS! So the sign of an upreared angry snake, as S, could well indicate the animal. When children reach the age of learning to read the first kind of book put before them generally has a picture of a horse or a bird or whatever the object, beside the written word. When mankind was a child, as it were, he started achieving language pictographically. First came the pictograph. Then he got to the next stage, the ideograph, which was a fairly crude attempt to represent thoughts and actions and even inner experiences. But this was as yet in no way level with the attainment of oral speech, not a true written language corresponding to oral language. Another momentous leap forward came when the ideographs became phonographs, that is to say when the written characters became representations, not of things that are seen, but of sounds that are heard. Sound was thus frozen on the

143

page. The flow was fixed. The flux was overcome. The oral sounds when uttered would perish, but their written representatives would remain. Man could now grapple with time. It need never flow past him any more. What was said or done on Monday, no longer at the mercy of memory, could be put on record and be available not only on Saturday but on any Saturday for years to come. A man sitting at his table writing his diary is really engaged in arresting the flow of time, and putting events into boxes which he keeps in his study.

It was not easy for the force of life, the elemental urge, to forge such an instrument. There have been races, many races we may suppose, who never managed to make the step from oral to written language. We catch sight of them in Frazer's *Golden Bough*. But this failure does not necessarily imply a poverty of oral speech. Professor Wilson mentions the case of the Cree Indians who lived for 'untold generations upon the plains of Western Canada, and had no written language prior to 1841; yet their oral language was as highly developed in its grammar and syntax as the English or any other modern civilized language.' But because of their lack of script they remained in a vanishing present world, cut off from their own history and achievements.

As we have seen, when man broke into consciousness he was then in possession of two worlds – the actual one, and his mental image of it. In the actual world everything is transitory; even the everlasting hills are not everlasting. Still less so are the achievements and history of man. They must pass away save what can be salvaged in the confusions of song and fable.

The invention of the written word changed all this. Everything done or said could now remain for ever in man's second world. Recollection was turned into the hard substance of books, piled up on shelves, parcelled in the corridors of libraries, so that when we enter the British Museum Reading Room we have all round us the memory of mankind in concrete. 'Words are fleeting in pronunciation but permanent when written down,' said Bacon. Freed from the flux of time, knowledge is given house; and a most important part of knowledge is geography, indeed more vital to us than history, a map being a fascinating form of written speech. We love maps because through them we know where we are and where other things are, we can never be lost any more. When man first became conscious his mind was a blank with regard to geography, and indeed he didn't know where he was until late in the Middle Ages. Now people have a map of the world in their heads. Before the written word a man in Cornwall had only the vaguest notion, if any notion at all, of the whereabouts of a man in Scotland. Now a fascimile of the outer world is lodged in every educated man's inner world. Perhaps we do not sufficiently acknowledge the wonder and significance of this unless we are like Helen Keller, who roamed so freely in the kingdom of her mind. Mr and Mrs Jones walk down the High Street to shop at the supermarket with their baby and their dog. Neither the dog nor the baby have any such facsimile in their heads. Mr and Mrs Jones may not be terribly well educated, but they carry with them into the shop a pretty clear outline of the geography of the British Isles. And of Europe for that matter. The Atlantic Ocean too. Also the Americas, and the wild washing of the Pacific waves. And if this is true of Mr and Mrs Jones, imagine the extent of this second world, this inner world, of the scholar with his freedom to roam in realms flung far and wide. Think of the historian inducing the past to render up its hoarded heaps; of the anthropologist adventuring back into the Stone Age; of the archaeologist digging up the artefacts that yield the secrets of yesteryear; of the astrophysicist interpreting the rays of light as if translating from a script; of the geologist peering down into the backward abysm of ages to the last table and the first fossils, so that the mighty mind, indifferent to the conditions of time, can read the pages of the flinty books.

There is nothing figurative or exaggerated about this, it is simple fact. The Elemental Urge, the Godhead if you like, goes forging on and on until its forgery at last comes up with Mind, another bit of earth in yet another guise, earth now looking upon earth, and vocally casting its unearthly spell abroad.

Publishers and editors continue to be astonished at the number of people who want to write. And there are thousands more, unknown to them, who say 'I want to write!' and who try their hand, and are less easily discouraged than the aspiring painter or musician. For is not a letter, writing? Is not gossip transferred to paper, writing? This powerful, almost irresistible urge in so many is due to the special nature of the medium, its special place among the arts in its defiance of time and its capacity for omnipresence. 'The Intellect can raise,' Wordsworth inscribed on a stone at Coleorton in 1811, 'from airy words alone, a Pile that ne'er decays.' And Tom Paine, in his *The Age of Reason*, wrote, 'Statues of brass or marble will perish, and statues made in imitation of them are not the same statues, nor the same workmanship, any more than a copy of a picture is the same picture. But print and reprint a thought a thousand times over, and with material of any kind, carve it in wood or engrave it on stone, the thought is eternally and identically the same thought in every case. It has a capacity of unimpaired existence, unaffected by change of matter, and is essentially distinct and of a nature different from everything else we know of or can conceive.'

It is free from the bondage of time. John Keats thought that such bondage could sometimes be regarded as a marvellous gain. He was inspired by the Grecian Urn because the figures inscribed thereon *were* fixed in time. 'Fair youth beneath the trees, thou canst not leave Thy song, nor ever can those trees be bare.'

> Ah, happy happy boughs! that cannot shed
>   Your leaves, nor ever bid the Spring adieu;
> And happy melodist, unwearied,
>   For ever piping songs for ever new;

More happy love! more happy, happy love!
For ever warm and still to be enjoyed,
For ever panting and for ever young.

Yes, there was a case to be made out here for the bondage, with no movement into the revenges of time. Even so, as Paine pointed out, there was only this one urn to carry the message, and even if we overlook the fact that an imitation is not the same thing, an urn cannot be published like a book and posted all over the world. There is only one Parthenon, and it is as irremovable as it is irreplaceable. Cleopatra's Needle on the Embankment in London remains the most sensational and poignant example of a special postal delivery – if we except the message sent, according to Herodotus, by Histiaeus to Aristagoras written on the shaven skull of a slave.

But words can be cast abroad over all the world, and such omnipresent preservation is an irresistible attraction. 'Not marble, nor the gilded monuments/Of princes shall outlive this powerful rhyme,' said Shakespeare. No problem there. But how about beauty? How is Helen of Troy to remain present and continue to launch those ships? How is beauty to be saved against the furious assaults of Time, he asked, since not even brass nor stone nor water could escape mortality? 'How with this rage shall beauty hold a plea,/Whose action is no stronger than a flower?' How could it possibly be preserved 'Against the wreckful siege of battering days,/When rocks impregnable are not so stout,/Nor gates of steel so strong, but Time decays?' A fearful meditation indeed! How could the beauty of his love be saved? Where could 'Time's best jewel from Time's chest lie hid?' The answer was at hand; for even as the pen of the master moved on the paper, he showed how it could best be done – with ink.

VI Memory

YET, SPEECH AND WRITING, which have made us into such a

communicating species, would be of little avail were it not for the possession of that astonishing faculty which we call memory. Nothing is harder to explain. 'When memory raises its head the sciences are dismayed, and fly before it,' writes Macneile Dixon. 'I look for a moment at a ship upon the sea and then turn away. I can still, however, if I wish, see it in the mind's eye. Where in the interval has been the picture, where is it now, how do I retain and recall it, perhaps months later? You may ask, but you will not be answered.'

Personally I derive satisfaction from something being inexplicable. I like mysteries that elude scientists, and the lack of certainty about this matter is its most pleasing aspect. I know that seven times eight makes fifty-six. But what is seven times nine? I have forgotten! I search my mind, it eludes me, and I have to make a roundabout calculation. I have forgotten my friend's name just as I am going to introduce him to someone; it is on the tip of my tongue, it is rolling about in my head just out of reach – how maddening. How can I have forgotten it? What is the process of forgetting? It is as strange as the process of remembering. I pass by a certain place, I detect a certain smell, and at once a flood of memories pour into my mind. Where have they come from? Where waiting? Tucked away in a millioned arrangement of minute shelves in my brain? But they are not substance, and need no shelves or drawers. Nevertheless they exist, it seems, inside me, and can come out or stay in, just as they please. We are told that the mind holds all that it has heard, learnt, experienced. It is all imprinted there as on a sort of *tabula rasa* of incredible accommodation. Indeed, there are some people who cannot forget, they are afflicted with the disease of over-recollection. Some simple men are plagued with being unable to forget a single sports result. Others are mentally distracted by being unable to learn how to forget, and the Russian psychologist, Alexander Luria, described a man whose greatest difficulty was to forget the flow of trivia which entered his mind; for the stream of consciousness *is* a stream, not to be noted save by men and women of genius, and even they must tread carefully before handing it on. The Argentinian writer, Jorge Luis Borges, describes in fiction the fate of a boy who having fallen from his horse forgot how to forget and was drowned in the stream of forgetlessness: 'I have more memories in myself alone than all men have had since the world was a world.'

Physiology can tell us about which centre in the brain is associated with memory, and while we get no nearer to an understanding of the problem of recollection, we can affect it. Take the case of 'Henry M.', as designated by Dr Colin Blakemore. When this man was young he had been operated on for severe epilepsy, that chronic disease of the nervous system which is characterized by convulsions. The surgeon decided to remove that portion of the cortex which he deemed was causing the seizures: the part which is called the hippocampus, just below the temporal lobes of the cerebral hemispheres. With certain patients the hippocampus had previously been removed with very satisfactory results. But the surgeon, regarding Henry's case as particularly severe, destroyed his hippocampus on both sides. This led to disastrous results. Henry's memory was destroyed at the same time. He is still alive, I believe, and in good health, and is apologetic about his predicament. For he lives only in the present. Every moment is fresh. He does not know what day it is, or what year, or when he was born. His dearest friends are utter strangers to him; each time he meets them he fails to recognize them. If one of them dies he experiences fresh grief each time he hears of it. Every day is alone by itself. He recollects no sorrow – and no joy. His case is not akin to that of amnesia which is a temporary block, when a person does not know who he is or where he came from, and remains in search of himself until he recovers his memory. But in the case of Henry M., the surgeon in removing his hippocampus had removed his memory.

Is memory then the hippocampus? Can you hold a hippocampus in your hand and say, here is a piece of memory?

Of course not. It is only the physiological agent of the mystery. For a mystery it remains. In her inspiring book, *Man on Earth*, Jacquetta Hawkes writes: 'If the temporal lobe is exposed and electrically stimulated its owner will be beset with vivid memories of days gone by, perhaps scenes from childhood and youth. How this substance, soft like tooth-paste, of the nerve cells, and little threads of nerve fibre running between them, can hold the impressions of a lifetime, and perhaps of previous lifetimes, and then project them once more like magic lantern slides showing a girl in a spring orchard, a boy riding with his mother in a carriage, is hard indeed to comprehend. Hard enough to have baffled the most formidable scientific minds of our time, the brains belonging to such great names as Sherrington, Adrian, and Head.'

Meanwhile most scientific writing on memory is the *ne plus ultra* of description without explanation. All the talk of 'coded messages', and of DNA and RNA, and arguments as to what kind of molecules could possibly qualify as the repository, leave us with the same mystery as to what memory *is*. Its absence or its presence does not seem subject to physiological inspection. This is not surprising to an ordinary person like myself. I cannot easily conceive how it could ever be explained why Mr Jones can scarcely remember a telephone number, while Macaulay could recall a page of print after a quick look, and recite to himself by way of relaxation the whole of *Paradise Lost* when crossing the Irish Sea. And if we really wish to explore memory in terms of an enormous mine (irrespective of what certain molecules have to do with it), we can read Coleridge's *Kubla Khan* and *The Ancient Mariner*, and then Livingstone-Lowndes' book called *The Road to Xanadu* in which he explores the empire of Coleridge's mind where lay hidden vast stores of reading culled from remote sources which came to the surface when he composed those poems. Physiological expertise can do little for us here. We can enjoy the pure mystery of memory unimpeded by the reductions of analysis.

150

For my part, I like to think how memory can take the place of the outer world if we so desire. Even while on a London bus. 'As I sat inside that crowded bus,' wrote Logan Pearsall Smith, 'so sad, so incredible and sordid seemed the fat face of the woman opposite me, that I interposed the thought of Kilimanjaro, that highest mountain in Africa, between us; the grassy slopes and green Arcadian realms of negro kings from which its dark cone rises; the immense, dim, elephant-haunted forests which clothe its flanks, and above, the white crown of snow, freezing in eternal isolation over the palm trees and deserts of the African Equator.'

We Sleep

🦋

I Sleeping

IT WOULD NOT be possible for our man in the field to remain standing there permanently. He would get tired. He must be allowed to lie down at intervals and sleep.

To sleep. It is a usual occurrence, and yet so strange that if it were not usual we would think it astonishing. We do note from time to time that the ordinary is extraordinary, but we continually regard the extraordinary as ordinary. To be able to die without dying is almost a definition of sleep. 'He is not dead, but sleepeth,' said Jesus looking down upon the unconscious Lazarus. It is not surprising that in olden times the spirit of the sleeper was thought to leave his body and to roam in realms unknown to the living man. 'There was a widespread belief in the ancient world,' says Dr Colin Blakemore, 'that sleep was a time of communication with the gods, a time when the spirit left the body to wander alone; and its experiences were the dreams of the night. The ancient Chinese would never rouse a sleeper hastily, in case his spirit did not have time to re-enter his body.' 'Can I see So-and-So?' someone may ask. 'I'm afraid not, for he is asleep,' may be the reply. It is sufficient; the enquirer does not persist, for it is taken as quite in order. Had the reply been framed: 'I'm afraid not, for he is unconscious,' or 'his spirit is absent,' or 'he is dead without dying,' the enquirer would not really have the right to be startled. Anyway, we do feel a trifle startled, not to say alarmed, when someone who is asleep does not lie down but walks about. It is a daunting sight to see someone walking or speaking or writing without being conscious of it. A ghost could be less frightening than the insensate actions of a sleep-walking person whose mind is distempered with anxiety or corrupted by sin. 'Out damned spot! out, I say!' owes its lasting dramatic power to the reality upon which the poet's imagination was based.

What happens when we sleep? There is, as yet, only a physiological answer to this. The necessity of sleep arises because the exercise of our bodily functions is in itself destructive of the tissues of the organs which minister to them, so if the waste produced from their action were not duly repaired they would speedily become unfit for further use; and it is upon the nutritive regeneration of the tissues which takes place during true healthy sleep that its refreshing power depends. Indeed that is sufficiently obvious as a good working description of sleep, seeing how worn out we soon become if deprived of it, and that a man will die of lack of sleep quicker than without food, though five hours a night is quite enough for many people, while nine is essential for others. In America they have gone in for experiments called 'wakeathons' to examine the effect of continuous imposed sleeplessness. The longest period was achieved by Peter Tripp who managed eight days, but not before he had experienced deplorable delusions and hallucinations. Since sleep does not necessarily overcome us by slow degrees, but sometimes swiftly snatches us into oblivion, it requires a great effort of will to keep awake when necessary. All the members of Scott's fatal expedition to the South Pole knew that sleep was akin to a sentence of death as they drew near to their last post. The will-power by which Lindbergh kept awake during his Atlantic flight entitled him to all the adulation he received.

The electrical activity of the tens of millions of nerve cells of which the cerebral cortex is composed can be recorded through the skull by an electroencephalograph, which is the most recent of modern instruments for the measurement of electrical activity. Never enamoured of capitals indicative of objects or organizations, I am yet glad that the electroencephalograph now qualifies as EEG. This EEG, by means of sensitive electrodes placed at given points on the scalp, can read the electrical activity, amplify the signals and record them on moving paper. With the aid of EEG the oscillations of a person who is awake and alert are shown to be fast and irregular; when drowsy the waves are slower and have been named alpha rhythm; on going to sleep slow waves alternate with bursts of high frequency known as sleep spindles. The EEG also shows that the basic rhythm varies with age from infancy, childhood, puberty and maturity.

In short, what we might have assumed in a vague way is told to us in an elaborate and accurate way by EEG, while that which was obscure before, such as the relation of consciousness to sleep, is relegated to deeper darkness. Writers on sleep sometimes tend towards obviousness, tautology, superficial assertion, evasion, and information masquerading as explanation – a notable exception being Anthony Smith who while being immensely factual and knowledgeable, is never evasive, and remarks in summary: 'Sleep is a vital need but its reasons are quite unknown. Dream sleep is equally vital, and equally incomprehensible.' Still, some medical writers have come up with Enkephalin, a peptone molecule in the brain, which builds up into an opiate against pain. Legendre and Pieron, two pioneering Frenchmen, claimed that this peptone can become so definite a chemical fluid that if transfused into the ventricles of another animal the latter will fall asleep. That was sixty years ago, and now it seems that Dr John Pappenheimer of the Harvard Medical School has come upon Enkephalin as 'factor S' in his studies. This substance when isolated from the cerebral fluid of sleepy goats and transfused will greatly enhance the sleep of rats.

All honour to Legendre and Pieron with their Enkephalin, and to Pappenheimer with his goats and rats, and to the inventors of the electroencephalograph. But I do wish the medical profession would be a little less modest about the achievements they have already so definitely made in the alleviation of insomnia. It is difficult to think of a greater benefit in modern times than that conferred upon humanity by sleeping pills. The scourge of insomnia has been largely overcome. Think of the last century in this respect. Take a single example. Thomas Carlyle couldn't sleep properly, throughout most of his life. The only time he really slept well was on the night before his Rectorial Address at Edinburgh. On that evening, when he and his friend John Tyndall were on their way to the city, staying at Lord Houghton's place, Tyndall decided that the only thing to do was to ride with Carlyle over the moors for *five hours*. They did this, returning with exhausted horses to an aggrieved groom. Even so, Carlyle said: 'I have no hope of sleep. I will come to your door at seven o'clock in the morning.' Yet he slept soundly till nine o'clock. Tyndall wrote the following comment: 'The change from the previous morning was astonishing. Never before or afterwards did I see Carlyle's face glow with such happiness. It was seraphic. I have often thought of it since. How in the case of a man possessing a range of life wide enough to embrace the demoniac and the godlike, a few hours sound sleep can lift him from the grovelling hell of the one into the serene heaven of the other! The question of sleep or sleeplessness hides many a tragedy. He looked at me with boundless blessedness in his eyes and voice: "My dear friend I am a totally new man; I have slept nine hours without once waking." ' It is certainly true that this single question of sleep or sleeplessness hides many a tragedy. The benefits bestowed by a sleeping-pill should not be underestimated. To have been able to select an ingredient which will affect what we may call 'the sleeping agent' within the cortex, is a very great thing.

Apart from this, it is worth mentioning that a number of people who find it difficult to sleep could achieve it quite easily without benefit of pills. Not by counting sheep. The person who thinks he can do it thus must have the brains of a sheep; the mere phrase is senseless. He could do it by taking a journey in the unboundaried kingdom of his mind. We have our inner world, that second world. We so seldom visit it, and roam about in it. After I have been abroad, or have had a particularly interesting day, I tend far too often to leave it all at the back of my mind and hurry on 'living in the present', as they say. If we cannot sleep, we can re-travel those journeys and re-visit those people. We are at liberty to do so, to return if we choose to the scenes of our childhood, even to the land of lost content. Of all the things we have been thinking about with regard to ourselves, there is nothing so strange as *memory*. 'When to the sessions of sweet silent thought/I summon up remembrance of things past,/'I sigh the lack of many a thing I sought,' and weep for 'precious friends hid in death's dateless night'. Yes indeed; but how much more there is than the agony of regret in the remembrance of things past. And how many scenes are there. We lower our bucket down into this well which we carry about with us, and bring up whatever we wish. We have been to a place. We treasure a place. We may never be able to visit it again. Yet we possess it for ever: it is there for ever inside our heads, and we can collect it and re-collect it as often as we please. There is a lake. At this moment I call it before my inner eye. Its name is Lough Dan, lost in the Wicklow mountains. I gaze upon a sequestered spot. It rests in solitude. There is no public way to it. A silver strand; a battered boat-house at the bottom of a rocky pile, and a ruined boat tethered by an ancient chain. The sun shines upon the sand where no one walks, and upon the water where no boats are seen. Here is a corner of earth walled from the world's woe. The paradise of solitude is here, and the voice of silence. I have left it, but not lost it. And if in seeking slumber I have summoned up its presence, I have now no need of sleep.

II Dreaming

IT IS NATURAL that as much attention has been given to dreams as to sleep, since they belong together. It seems to me that the writers are helpful in terms of analysis; that is to say when they point out that there are two principal kinds of sleep, the orthodox and the paradoxical. For some reason they give the term orthodox to sleep without dreams, and paradoxical to sleep with dreams. By breaking down sleep, with the aid of EEG, into different kinds, the experts are able to prove that dreaming is necessary and that a 'dreamless' sleep does not happen any more than sleeping like a log has any foundation in fact.

In his stimulating and informative Reith Lectures Dr Colin Blakemore assumes that we 'cherish dreams, the sole conscious product of sleep', and claims that from the soothsayer Artemidorus of Ephesus to Freud, pundits have capitalized upon the fascination of these products of the undisciplined mind for us. He goes on to say that 'the similarity between the untrustworthy image of the dream and waking experience has posed enormous problems for the philosopher,' and he quotes the third-century Taoist, Chang Tsu, as one of the first to express his doubts and fears. One night he experienced a very vivid dream that he was a butterfly happily flying here and there. When he suddenly woke up and found that he was really Chang Tsu, he was disconcerted. 'Did Chang Tsu dream he was a butterfly?' he mused, 'or did the butterfly dream that it was Chang Tsu?'

Personally I do not think that Chang Tsu was full of doubts and fears about this. He was just having his fun. And I would contend that there is no similarity between the waking and sleeping experience, indeed few things could be less similar. As for the supposed 'fascination' of the subject of dreams, only stupid or learned people are not bored with the subject. People who talk to me about their dreams send me to sleep. My own have no interest for me save on the rare occasions when something nice is happening which, on awakening, I want to continue but cannot, and on other occasions when I am intensely glad to wake up and find that actually I am not lost or my car has not been stolen. But I find it very difficult to learn from the authorities why I dream more if I am too hot in bed, or too cold, and why if I have indigestion my dreams are so unpleasant. What has the stomach to do with this? Why should imperfect activity in the intestines transport me in a dream to a dungeon in Barcelona?

Learned men are interested in dreams because they hope to *interpret* them. This is where 'the enormous problems are posed' for Blakemore's philosophers. Everyone has heard of Freud's *Interpretation of Dreams*, and of the sexual gloss he gave to the interpretation, and of his axiom that dreams convey unconscious desires. He said much more than that, but the great unreading public has decided that these ideas compose the chief aspects of the book. And it is true that Freud's love of the absolute statement is absolutely absurd. To proclaim that behind the symbolism of dreams there lies ultimately a wish, and that a dream is wish-fulfilment, must leave even a great thinker open to derision. In his *The World of Dreams* Havelock Ellis wrote: 'Those who imagine that all dreaming is a symbolism which a single cypher will serve to interpret, must not be surprised if, however unjustly, they are thought to resemble those persons who claim to find in every page of Shakespeare a cypher revealing the authorship of Bacon.' Havelock Ellis held that it would be just as proper to advance the contrast-dream as opposed to the wish-dream theory; that is to say the dream in which characteristics emerge altogether opposed to the dreamer's own character and habits. He supplied a few examples from his own experience when on four consecutive nights he dreamed that he was the mayor of a large northern city about to take the chair at a local meeting of a Bible Society; a soldier in the heat of battle; and a young man meditating about the step of going on the stage as a comedian.

With Jung we enter a different domain; not clinical in the Freudian manner; not an area where 'philosophical problems are posed', but into a vast intuitive region – factually supported, let me emphasize, by the examination of 67,000 dreams of his patients. For him it was very much a question of personal experience and conviction. He regarded dreams in terms of inner voices coming to him from the unconscious with a message that would 'keep him on course', to use a phrase by Laurens van der Post who holds that the dream 'is part of the unconscious made accessible to our waking selves in sleep; a potential form as it were of conscious unconsciousness.' Jung saw dreams as the link between our conscious self and the psyche and as the regenerators of immemorial symbols. He was only concerned with what he called great dreams as opposed to the trivial everyday or everynight dreams about missing trains and so on; he connects great dreams with the collective unconscious, and he held that in some of the dreams the unconscious is seeking to give guidance or convey a warning to the conscious mind. My own grasp of matters in this field is somewhat limited, and as I am unwilling to write upon anything I imperfectly comprehend, I will say little more about Jung here in spite of the impropriety of brevity on so illustrious a name. But I would emphasize that Jung's dreams which came to him in terms of pre-vision had an appalling significance, for he prefigured both World Wars, forecasting the Second as starting in 1940.

It is only on rare occasions that I have glimpsed what Shakespeare meant by saying that 'we are such stuff/As dreams are made on, and our little life/Is rounded with a sleep.' I warm to the remark by Marcus Aurelius: 'When thou hast roused thyself from sleep, thou hast perceived that they were only dreams that troubled thee. Now in thy waking hours look at these things about thee as thou didst at thy dreams.' A learned friend of mine is fond of reminding me that though people say that they can't conceive of 'another life' they nevertheless live one in sleep, 'and a damned odd one it is'. I hope I shall be spared going to such a place. But I like to think of Bulwer Lytton who wrote a book in which a man spent a lifetime when his real world was his dream-world at night. He endured his waking part as best he could, and then every night returned to a consecutive life in his dreamland which was his

reality and his continuous delight. How many might cry:
Would that it could be so! 'What do we here?' asked Blake:

> What do we here
> In this land of unbelief and fear?
> The land of dreams is better far,
> Above the light of the Northern Star.

We Reproduce Ourselves

🐞

I   The sperm

FINALLY, there would not be much point if our man in the field were to remain all by himself. He might become lonely. There would be no one with whom he could share his view of things. It is necessary that he should be able to reproduce himself, or something of the kind. And, as we know, he does this. But he cannot do it alone. So at this point in my narrative I am compelled to bring onto the field another member of the species with a particular difference in physiology. Since it is common knowledge, I am not called upon to say what a woman is. And for the same reason there is no element of surprise in stating that reproduction is only possible if the man is able to pass a portion of himself into herself. He passes a seed to her, and she, combining it with a seed of her own, is able to produce a third person.

'Get rid of your miracles,' said Rousseau to the Churches, 'and the whole world will fall at the feet of Christ.' By this he meant that the miracles performed by Jesus were irrelevant to Christianity; that cures for physiological defects have no bearing upon Doctrine of any kind. And indeed it is true that 'the miracles' are a frightful red herring. No doubt Rousseau also had in mind the plain fact that we have quite enough miracles to go on with, and have no need of anything special or confusing of this kind.

If, looking round, we choose the miracle which stands out above all else, it is that of birth. You have a speck of a man's semen and a speck within a woman called an ovum: they get together and build up a thing as elaborate as a man who may be six foot tall, or a woman, perhaps like Helen of Troy. If I

161

were asked, before the event, faced with these two specks, whether such a claim is not rather extravagant, I would say, no it is not extravagant, it is impossible: you might just as well suppose that two specks of dust could combine to build St Paul's Cathedral. Yet we know that the one sperm and the one ovum do perform this miracle, and that this rules out of court any conceivable theory of mechanism.

The facts concerning the process are, I think, worthy of a brief statement. Taking the male first, we have the testicles in which spermatozoa originate. In the mass we discern these as semen. They were not individually observed before the discovery of the microscope in the seventeenth century, for they are so minute as to be invisible without powerful optical aid. But in nature the size of a thing bears no relation to its importance or its complexity. The sperm possesses a head, a neck, and a tail, the last being the longest part, about one five-hundreth of an inch. The head is not so large, but being more compact is easier to detect if we magnify it one hundred and sixty times. Within this head there lie chromosomes to the number of twenty-three, visible also through the microscope, and within these minute parcels lie their constituents, the invisible ultramicroscopic entities, the genes, which are responsible for the factor of inheritance. The number of spermatozoa spurted from the penis at the climactic moment of ejaculation is something in the nature of two hundred million. Yet only *one* of them is responsible for the fertilization of the woman. (In my youth I used to imagine that the more semen a man pumped into a woman, the more healthy would be the baby, and that that was why he was congratulated.)

The female equipment for fertilization consists of ovaries instead of testicles. The ovaries have two main functions: the production of ova, and the manufacture of hormones, which may be defined as chemical substances which have specific effects such as the development of genitalia, the growth of breasts, and so on. The ovary, by means of certain structures known as primordial follicles, which after puberty number some hundred thousand, produces one egg per month during the thirty to forty reproductive years of a woman's life. This egg possesses twenty-three chromosomes just as the male sperm does. And it is in search of this egg that the two hundred million spermatozoa after ejaculation, propelled by their tails, swim like fish in a race up the female vagina which is the canal leading to the uterus (synonym for womb). The winner of the race of the two hundred million will encounter, or fail to encounter, the egg. If it does so, then fertilization will follow. The twenty-three chromosomes in the sperm combining with the twenty-three chromosomes in the egg will produce a human being.

The reader will be aware that in writing the words fertilization will follow', I have explained nothing and have attempted only a brief description of the process. We are faced with the Act of Creation as inexplicable as the appearance of the primordial cell in the first waters of the world. We are faced with the genius of the genes which ever since Genesis from generation to generation have built up the human race. Not even visible to mortal eye, they have created eyes, and bodies, and minds which can brood upon the mystery of the inheritance of characteristics so continuously acquired that the great-grandson of the Duke of Wellington bears a striking resemblance to his illustrious ancestor. To be born is a strange occurrence, and even the least reflective person is constrained to wonder at times why he is here, and who he is. The great Pascal, a mathematical genius who at the age of twenty-four turned from science to religion, always carried a parchment sewn into the lining of his doublet, found at his death, which was a memorial to his mystic experience and a summary of what he meant by the words which he gives to Christ: 'Console thyself, for thou wouldst not seek Me if thou hadst not found Me.' He exclaimed that when he considered the infinite immensity of spaces of which he was ignorant, 'I stand terrified to see myself here rather than elsewhere, for there is not the slightest reason for the here rather than the elsewhere, or for the now rather than for some other time. Who put me here? By whose order and direction have this place and this time been allotted to me?' Indeed yes, these are the thoughts which should occur to us when we contemplate the chromosomes. 'In each family a few, a very few, out of legions of possible human beings come into existence,' says Macneile Dixon. 'They are, shall we say, the favoured few? Why were they, like ourselves, so singled out? And at what moment did this self of ours, so precious to us, this "I", this individual person attach itself to the chromosomes from which our bodies have sprung?'

Scientists may feel less inclined to ask such questions or to take so lofty a view of the human situation. They are more likely to tell us that they themselves could, without benefit of sexual intercourse, produce a living person – or at least expect to be able to do so. One can see that this would be theoretically, and even practically, possible; and since, as I say, going to extremes is a marked characteristic of our species, it may yet be done. It would be entirely unnecessary, entirely futile; but it would be done just for fun as a 'significant scientific experiment'. Eventually a spermatozoon and an ovum could be isolated and conditions for fertilization made possible, resulting in the birth of a human being as complete as you and I. But this would not mean that the scientists had produced life, since that lay in the isolated sperm and ovum with their appropriate chromosomes and which they have no more chance of manufacturing than the primordial cell. Thus the whole thing would have been rather a waste of time.

II Fertilization and birth

BUT LET US LEAVE the philosophers and the scientists and get back to the Becoming of the Being. The race was to the swift. The winner on the two hundred million sperm-course has met the ovum, and fertilization has taken place resulting in the appearance of a pin-point of an embryo. That which was not, now is. And that which now is contains within itself the power of becoming a human being, with all its organs complete, brain, heart and lungs, a vascular and a nervous system, liver and kidneys, bones, muscles and limbs; it can develop the eye, the ear, the will, the emotions, the thoughts that make a man. There it is now within the woman. But where exactly? At a place which the embryologists call the ovarian end of the Fallopian tube. It does not stay there. It must find a better place. It seeks suitable residence in which to grow. It has already started to grow. It began with two cells. After the third day of its creation it has twelve cells, and will eventually accumulate two hundred million, while its weight will increase five hundred million times. Thus the cells of the embryo are multiplying apace and it must find accommodation where it can grow in comfort. It is still a floating object with as yet no harbour in sight. At last it comes to the uterus, the womb, which at this stage is about the size of a fist. Now it attaches itself to the side of the cavity like a minute limpet or sea-anemone. It plants itself there, the process being known as implantation. And to some extent it behaves like a plant; for while it puts down a root in this place it grows a kind of stalk which goes out into the womb. The bud at the end of this stalk is the foetus which will become the baby just as the bud at the end of a rose-twig will become a rose. This stalk is known as the umbilical cord linking the foetus with its root in the uterus. Thus the embryo goes out into the womb, floating in its salty and sugary fluid, in order that it may have room to expand. The cord is not really a cord in the sense of a cable thrown out as an anchor; it is a pipe, a two-way pipe, by means of which the foetus will receive food and discard waste.

It will need blood and it will need oxygen. How can this new thing, a foreign body, enter into communication with the vascular system of the woman in whose body it now resides? It can do so by means of an intermediary vascular system which has been prepared in the uterus, called the placenta. The uterus has two or three layers, the inner of which is responsible for becoming the placenta. Each month this layer grows in anticipation of being used for nurture. If there is no conception then this lining which constitutes the menses is cast away, the process known as menstruation – regarded as a curse if conception is desired and as a blessing if it is feared. In summary the placenta may be defined as a vascular organ within the uterus connected to the foetus by the umbilical cord, and serving as the structure through which it receives nourishment from and eliminates waste matter into the vascular system of the mother.

The womb was a small place when the embryo entered it. No matter; for while the cells multiply the uterus enlarges, while the foetus swells its house expands, so that as week by week and month by month go by and the embryo becomes ever bigger, taking upon itself the semblance of a living creature, the space in which it has taken up its abode inside the woman becomes so much more spacious that her figure is seen to alter in some cases as much as between a straight line and the figure eight. One might suppose that, however minute, the embryo would appear from the beginning as a rudimentary human being. But at first it is obliged to repeat in itself in rough summary the events of hundreds of millions of years from worms through fishes and amphibians to mammals, so that at early stages it is indistinguishable from a potential bird or fish, and during that time the mother might more fairly be said to be carrying a fish than a baby. However, in due course the human characteristics assume preponderance, and the foetus acquires all the attributes of humanity. The time comes when the woman labours to expel this foreign body which is within her body. At a later date, at moments of passionate or aggrieved possessiveness, she may speak of the child as 'her own flesh and blood'; but always the son or the daughter will be alienated by this remark as vulgar and unseemly as well as untrue, for there is no exclusive ownership in this matter, and each son and each daughter know that they belong to themselves and primarily to the primordial cell: 'Muse on her, do not forget our Common Mother . . .'

The men who interminably talk sex and smut are silent about childbirth. *These* facts of life are too intimately related to their theme – so drastic a result from so often a perfunctory act. Indeed, smut-merriment has often puzzled me. It argues such a lack of imagination. Semen! Fancy joking about semen. It is spurted out, a grey, pale-milky fluid – and then the man changes. For he has parted with energy; the fluid is vitality made manifest; it swarms with life; it is population in the raw. And when it has left him, when that amount of force has drained from him, he feels the loss for twenty minutes or more than twenty days according to his cycle. This feeling of weakness and loss turns him against himself, so that he names his action Lust. The mildest line in Shakespeare's celebrated sonnet ('The expense of spirit in a waste of shame') is – 'Enjoyed no sooner than despised straight.' I once heard a beefy hulk of a chap say: 'Arter I dun it, arter I got rid of the stuff, I could throw 'er in the ditch.' Such reactions do not always prevail by any means, indeed often quite the reverse. But between Shakespeare's sonnet and the unseemly remark by that heartless fellow there is enough truth to make smutty jokes no laughing matter.

III Abortion

THE QUESTION IS continually raised nowadays as to whether the foetus should be considered as a human being. Actually it is legally classified as such after seven months of pregnancy. This would seem a reasonable ruling. We cannot really doubt that it is alive before it is actually born. If we need proof of this we have it in the rare but definite phenomenon known as *Vagitus uterinus*, when a cry comes from the uterus before birth. The most recent case occurred in Scotland in Inverness-shire.* A mother, with a previous record of difficulties in delivery, was five days overdue. The doctors decided that labour should be induced. They ruptured the membranes, and while the sequent liquid was withdrawn, three loud cries came from the foetus. These cries were heard by all present, including two doctors and three midwives, not to mention the mother herself. They were startled by this and thought that someone had inadvertently brought a baby into the room. But it was soon clear that the cries had come from the foetus, for they were heard again frequently for eleven hours. The doctors decided that there was nothing to be done but wait upon the event; after nineteen hours since the first cry from the foetus, the child was born, and immediately cried again as we all do when coming into the world. Everyone present was very relieved to hear it, and to find that both mother and child were unharmed. Such cases are rare, but they do occur from time to time. The cries may be loud and gasping, or they may be soft and whimpering as from some imprisoned soul in preternatural distress.

Thus there is no point in denying that when we kill a foetus, as in abortion, we are killing a living entity. This would seem a commendable thing to do if those responsible for the foetal-being are unable or unwilling to cope. To prevent its growth at an early stage would seem the best policy, and indeed it is pursued in most countries these days. In England and Wales there are one hundred thousand abortions a year; in America a million; in Denmark twenty thousand a year with a population of four million. In Japan during 1960 legal abortions outnumbered births, while in Hungary statistics claimed at one period the number of abortions to be one hundred and thirty-one as against one hundred births. Some Russian women even claim that having an abortion is like going to the hairdresser. Taking the world at large the figure for abortions has been given as thirty million a year. This

*See *The Body* by Anthony Smith.

168

would mean in ten years the absence of three hundred million unwanted children on the surface of the globe.

There are people who object to this. They speak about 'the sanctity of human life'. And they use the word 'murder'. They do not speak about the sanctity of *life*. It is Nature's way to pour out a prodigious abundance of progeny, sometimes even raining flesh and blood as in the case of locusts. Life scatters its seed like dust. The herring produces 40,000 eggs a year, the tapeworm a hundred million. The sources of life seem inexhaustible. Yet not really so, for when man appears on the scene and is determined to exterminate some of them, say the whale or the tiger or the buffalo or certain fishes and birds, there comes a time when the births fail to balance the murders. Man has nothing to fear from the animals with regard to his own numbers, for even epidemics caused by germs are largely under control. So now the immense fecundity of man can go unchecked; or if checked as in World Wars One and Two, can be as easily replaced as houses can be rebuilt. We can overrun the whole earth now if we choose, except those places which are uninhabitable or have been made so by ourselves. Few governments today consider this a good thing and measures against conception have been taken, but since they have proved inadequate the legality of early and safe abortion has seemed the most humane policy. While the statistic of thirty million may be incorrect, the number is significant.

But, as we see, there are people who agitate on behalf of over-population and declare that human life is sacred. They are more concerned with the eclipse of a foetus than the slaughter of thousands of people in ludicrous wars largely caused by over-population. They are fond of quantity. It is nothing to them that the foetus instead of being painlessly denied growth will be born to live in misery or die in famine. Sometimes I have received applications to contribute funds for the 'starving children of India'. It seems an unreasonable request. The Indians were, in a way, behaving like a man who, observing a large field, says: 'Not a bad area; a hundred of us could live here at peace with one another and ecologically at harmony with Nature: but I tell you what we'll do – we won't have a hundred people here, we will have a thousand.' It is like saying: 'Let's have life in order to destroy it; let's have abundance of people in order that they may be abundantly miserable.' There were periods in Chinese history when the inhabitants produced so many of themselves that they were always obliged to live at starvation level. At one time the population reached six thousand per square mile. Thus they sowed an extra crop, not of rice and corn, but of *themselves*. They could not harvest it, and were obliged to plough it in. Between 108 B.C. and A.D. 1911 there were one thousand eight hundred and twenty-eight famines in the course of which several millions perished. In the famine of 1920–21 the death roll was five hundred thousand, while twenty million were reduced to such hunger that they ate flower-seeds, poplar-buds, sawdust, thistles, leaf dust, elm bark, roots, and stones ground into an artificial flour as an aid to the digestion of withered leaves. What price 'the sanctity' of human life there? I prefer the word prodigality. 'I worship Him,' said Goethe, 'who has infused into the world such a power of production that, when only a millionth part of it comes out into life the world swarms with creatures to such a degree that war, pestilence, fire and water cannot prevail against them.'

IV   Potency

SINCE NATURE has made sexual desire so purposeful a function it is strange that there should be such a thing as homosexuality and lesbianism. It is not found among the animals who have never known how to refine and expand their pleasures in this way any more than in eating and drinking. In so far as the deviation has developed at all, it is surprising that it has not developed more, seeing that there is so much antagonism between the sexes.

Homosexuals are supposed to be less potent than other men. In a great many cases this cannot possibly be true. To perform mighty works, to put out great force, whether in physical or mental fields, calls for more than normal physiological energy. And that energy, that potency, whether deviated, perverted, frustrated, sublimated, or normal, must be, *au fond*, sexual. I speak under correction, and my terms may not be as exact as they should be, but surely the power, the force of life lies in the genitalia. The whole question of potency, as it seems to me, is troubled and vexed by narrowness of interpretation, and even those who have read the prominent sexologists and the Kinsey Report and all the rest of it may end up with knowing too much or too little, in a quite simple way. It is with feelings akin to indignation that I hear people refer to the 'impotence' of Carlyle and Ruskin, for instance, whose force swept across the British Isles and far beyond in the nineteenth century. I sometimes hear the same thing said about Bernard Shaw, for several goodish reasons. In the first place he once told Cecil Chesterton, who had asked whether he was a puritan in practice, that the sexual act was to him monstrous and indecent and that he could not understand how any self-respecting man and woman could face each other in the daylight after spending the night together. And when St John Ervine, on reading *Back to Methuselah*, objected to the wry face made by Eve when the serpent whispers to her the secret of reproduction, Shaw said that what made the God of the Eden legend incredible was 'His deliberate combination of the reproductive with the excretory organs, and consequently of love with shame.' A bit insane, you think? Indeed yes, for Shaw was a very sane man, and there is nothing more extreme than the insanity of the sane. I am sure that any man who displays great energy, who puts out great vital force and accomplishes mighty things in any field of life, has great sexual strength, whether that vitality is elevated, curbed, botched, or diverged. I do not mean that the potency need take a diverted path (think of Tolstoy or Wells); but to talk

171

about the impotence of a Ruskin or a Carlyle as against the wonderful potency of every Tom, Dick and Harry filling a Wembley Stadium is to narrow the word 'impotent' down in a monstrous manner. Whatever we may think of the wind-baggery of Carlyle and Ruskin, their wind blew across the land in a most potent way. As for Shaw he was no more impotent in the conventional sense than H. G. Wells, as his written testimony on the matter to Frank Harris makes perfectly clear; but he sublimated the force when he married a woman who detested sexual intercourse. (It is impossible to fathom this extraordinary marriage, and very little light has so far been thrown upon it; their correspondence is strangely unrevealing.)

The fact that so enormous an emphasis is given today to copulation can, without impropriety, be traced to Freud. While no one can seriously fault Freud in his assertion that human energy is essentially libidinous, it does not follow that human motivations are always libidinous. One might just as well assert, to use Mr W. V. Butler's apt phrase, 'that because cars run on petrol, the only place a motorist ever wants to get to is a garage.' But today people are so enslaved by sex, Bernard Shaw argued, 'that a celibate appears to them as a sort of monster. They forget that not only whole priesthoods, official and unofficial, from Paul to Carlyle and Ruskin, have defied the tyranny of sex, but immense numbers of ordinary citizens of both sexes have, whether voluntarily or under pressure of circumstances easily surmountable, saved their energies for less primitive activities.'

That there are under-sexed and over-sexed men is a matter of common knowledge and experience, but I think genius demands great physiological strength in the sexual sphere. I mean the power needed lies *there*: think of the Herculean power that a man often feels when he is aware of the semen arising in him; surely it is significant that he should have this feeling (concomitant indeed also with violence, sadism and masochism). Genius depends upon the strength, the energy, being there in the person. Whenever the subject comes up about the question of the lack of genius as displayed by women compared with men, the old chestnut is always hauled out that women have not had the opportunity, their bondage to home and children being the reason. That holds good with regard to talent, but not to genius. When women are as free as possible – which can never be as much as men – then there will be a lot more talent deployed by them than at present, not only in the arts, but perhaps especially in law, medicine, organization, politics, education. But not any more genius. Obstacles do not impede genius, it even thrives on them; it is talent that needs encouragement and freedom. I do not think that any amount of extra time and freedom could make possible a woman Shakespeare or Beethoven or Plato or Dickens or Kant or Turner or Shaw or Picasso or Cobbett or Michelangelo or Leonardo da Vinci or Napoleon or Winston Churchill or Wordsworth or Tolstoy. Random names up to a point, you may prefer others; but I think the reason for this is physiological. Women, we must remember, produce mankind and are rather good at looking after mankind. Men cannot do the first and are not much good at the second. There is in some of them a greater energy than women can ever command, sometimes amounting to superhuman force. Surely the root here is physiological, in the male hormones and genitals. It is *not* a question of freedom from chores. In three of the arts women are just as good as men: in acting, in singing, and in dancing. The handicap of female bondage to chores makes success in these fields *more difficult* than for writers, painters, or musicians. And why are they as good as men in acting, singing and dancing? I think it is because those arts do not demand the sheer creative energy called for in the other arts.

Judging from what the psychiatrists tell us, with the interesting support of those admirable women journalists who answer letters in a weekly column, a great many young men today are in a panic about their sexual potency. For the mores of the day exercise such a powerful influence. There was the dreadful Unpermissive Society of the last century with countless victims sacrificed upon its smoking altars. There is the Permissive Society of our times with its own victims and insistent 'make-love-to-me' demands. In spite of all the books and talk on sex, young men and women start from scratch, victims of the current attitude. A man will fail to take into account a girl's menstrual period (as later he may fail to make allowance for the menopause), while a girl will be amazed to find a man, after the sexual act, in a mood of sadness, or indifference, perhaps merely saying after a suitable interval 'As it is rather late, maybe I should be going now.' The force has gone out of him and he wants to think of something else. But as current thought does not allow for any degree of melancholy just then, the girl doesn't understand it. The attitude of the age in E. M. Forster's day made it possible for him to remain in ignorance of the sexual act till he was twenty-nine. And when Daphne du Maurier grasped it at the age of nineteen she exclaimed: 'What an extraordinary thing for people to want to do!' I didn't gather it till I was in the army, and it gave me quite a turn. How green we must appear to the multitude of permissives. And how they worry about sex! Perhaps we should all worry much less and not be 'put upon' by the times. Recently I read of a tribe in one of the islands of the Indian Ocean where everyone seemed to have a wonderful time sexually, the men performing prodigious feats; subsequently I read about a tribe in another part of the world who often forgot all about sex for six months on end yet were very cheerful, it appeared. There is not much to be cheerful about on this matter in the modern world. For added to the Lord of Misrule reigning over all lands there is the *new censorship*. The idea is that nothing should be censored. In the resultant free-for-all the really low stuff gets full support from those who can calculate on financial gain; and avant-garde work can still exist by virtue of élitist support. But family entertainment, not being avant-garde, can find no home: it is virtually suppressed, censored. Because there is no censorship for that which is low, there is censorship for that which is higher.

Epilogue

'A HUMAN BEING is a very complicated physical mechanism and nothing more.' Such were the words with which Professor J. J. C. Smart, having taken all the considerations into account, ended his lecture in a Series of philosophic talks on the radio in 1976.

People are inclined to become a little depressed by such statements made by the new materialists. I didn't actually realize that they existed as such until the other day when I heard Sir Alfred Ayer declare that he was 'suspicious of the claims of the new materialists'. I remember in my youth G. D. H. Cole writing angrily: 'I am not prepared to have my universe governed by chance.' And I recall how J. W. N. Sullivan in his *Galileo, or the Tyranny of Science* represented poets as being 'depressed by the Iron Laws of the universe,' and referring to certain scientists as 'men who must have been theory-mad, soberly maintaining that little particles of matter wandering purposelessly in space and time produced our minds, our hope and fears, the scent of the rose, the colours of the sunset, the songs of the birds, and our knowledge of the little particles themselves.' He got quite worked up about it. But even in those days I remained reasonably cheerful at such antics, saying to myself: Well done atoms, by God! What a fine job you have accomplished. If you can do that you can do anything. And there was Cole not wanting to be governed by chance. But since that government is so very striking, I said to myself, good luck to Chance, let us praise him, bow the knee, and call him god if he likes the term. The same with Professor Smart and his mechanical human being: it is with more than

176

perfunctory plaudits that I salute mechanism and subscribe myself his admirer, indeed his humble and obedient servant. 'Have faith,' said Edward Carpenter, still a bright particular star in the firmament of my heroes. 'Have faith. If that which rules the universe were alien to your soul, then nothing could mend your state – there were nothing left but to fold your hands and be damned everlastingly. But since it is not so – why, what can you wish for more? – all things are given into your hands.'

I open my Epilogue with these considerations because I have formed the impression that for every one person who is interested in physics or physiology, a hundred, a thousand, are interested in metaphysics. They may not call it that; they may prefer the term religion, or philosophy of life, or search for truth, or the meaning of existence, or just the question of personal identity. And they are not very keen on being told that they are just a bunch of chemicals. I feel that any reader who has been kind enough to join me as far as this will not wish me to sit on the fence here. If my opening remarks are not sufficiently lucid I will hope to clear them up before the end of the Epilogue.

It is instructive to glance at the philosophic approach to reality as exemplified by some of the greatest names, especially in relation to perception, which is so relevant to our theme. There was the Englishman, John Locke, who turned the weapons of reason upon reason itself. How does knowledge arise? Have we innate ideas, such as right and wrong, ideas about God which are inherent in the mind prior to all experience? The theologians listened carefully, for they would have very much favoured a positive ruling on this. They were disappointed. Locke held that all our knowledge comes from experience through our senses; in fact that there was nothing in the mind that was not first in the senses. The mind is a *tabula rasa* upon which sense-experience is written in hundreds of ways until sensation begets memory and memory begets ideas. Thus, since only material things can affect our sense, we know nothing but matter and must accept a materialistic philosophy. If sensations are the stuff of thought, matter must be the material of mind.

But Locke had not reckoned with the Irishman, George Berkeley, who with a fine stroke of Gaelic audacity turned the tables on him. Locke, he claimed, had really only proved that matter does not exist except as a form of mind. We know no such thing as matter. Locke tells us that all our knowledge derives from sensation. Therefore all our knowledge of anything is merely our sensations of it. An object is only a bundle of interpreted sensations. Is a mutton chop not substantial? Oh no, said Berkeley, your chop is at first nothing but a congeries of sight, smell and touch, and then of taste, and then of internal comfort and warmth. Its reality for you is not in its materiality but in the sensations that come from it. Is not the hammer that hits your thumb a substantial object? No, said Berkeley, the reality of the hammer is in the sensations that come from the thumb. If you had no senses the hammer would have no true existence for you; if it struck your thumb as a kind of hammer-ghost you would not wince. It also is only a bundle of sensations or a bundle of memories; it is a condition of the mind. Materialism has no part to play.

Such an approach may seem far-fetched, but it made considerable impact and engaged respectful attention. Dr Johnson struck his foot against a large stone, till he rebounded from it saying: 'I refute it *thus*.' In the next century Leo Tolstoy, under the same influence, thought that possibly if he withdrew his attention the world about him might disappear, and he would turn round suddenly to see if he could catch sight of nothingness; once when he was going fishing he became so lost in thought 'thinking that he was thinking about what he was thinking about,' that he helped himself from a jar of worms in mistake for a piece of bread. A current wit declared: 'Since Berkeley says there is no matter, 'tis no matter what he says.'

But Berkeley had not reckoned on the Scotsman, David Hume, who came forward with the great book called *Treatise on Human Nature*, and turned the tables on *him*. We know the mind, said Hume, in the same way as we know matter – that is, by perception, though in this case it is internal. We don't perceive any such entity as mind; we only sense distinct ideas, emotions, memories, feelings. The mind is not a substance, nor is it a box holding ideas, it is only an abstract name which we give to our series of ideas, *they are* the mind, there is no detectable 'soul' behind the processes of thought. In his *Treatise*, which is a very powerful book, Hume was unable to find any nexus between cause and effect, and absolutely no place for the self. The book created a great stir, for it seemed that Hume had destroyed mind as Berkeley had destroyed matter. Another wit of the day suggested that the controversy be now abandoned, saying:'No matter, never mind.'

But neither Locke, Berkeley nor Hume had reckoned on the advent of the German, Immanuel Kant, who in his *Critique of Pure Reason* brought forward what still seems a more sensible analysis of sensation. The mind, he said, was neither a *tabula rasa* nor an abstract name for a series of mental states; it is an active organ which transforms the multiplicity of experience into the ordered unity of thought. What is meant by sensations and perceptions? A sensation is just the awareness of an experience such as a sound brought to the brain along the afferent nerves. A thousand varied sensations are sent by the afferent nerves to the brain – a medley of messengers all calling for attention. Plato had called them 'the rabble of the senses'. Left to themselves they would remain a rabble and could never be grouped into perceptions. Left to themselves, as Locke, Berkeley and Hume were content to leave them, they would never get transmuted, they would have no meaning, power or purpose. 'As readily might the messages,' writes Will Durant, 'brought to a general from a thousand sectors of the battle-line weave themselves unaided into comprehension and command. No; there is a law-giver for this mob, a directing and co-ordinating power that does not merely receive but takes these atoms of sensation and moulds them into sense.' This moulder that makes sense of sensations is the mind.

There is stimulation to be derived from studying these great minds on mind, not only because of the vast professionalism of their assault but for the comedy in the condemnation of their opponents. But I do not always find it easy to follow the German metaphysical idealism, and have especial difficulty with Hegel who ironically inspired Karl Marx to produce his *materialistic* conception of history. It seems that Hegel himself wished that he had been more widely understood; and I take comfort from his last words. According to Max Beer, as he lay dying and his disciples gathered round his bedside, it was seen that he wished to say something. They bent close to hear his words. 'No one has understood me,' he said. 'No one, except Michelet, has understood me.' Presently it appeared that he had something more to say. The disciples bent lower to listen to the last words that might fall from the lips of the Master. 'And even Michelet,' he said, 'did not understand me.'

All the same I swear by Kant. For there is so much in him, and he was such a prophet of creative evolution long before it was proposed with scientific chapter and verse. I have already reminded the reader of that theory in my piece on the birth of language – which made our man in the field different from all the rest of creation. It is no paltry theme. It tells how the force of life at last came up with consciousness so that nature could take a look at nature, a piece of earth behold the earth.

And more than the earth: a good deal of the universe as well. 'The distance of the nearest fixed star,' writes Bertrand Russell, 'is about twenty-five million million miles. The Milky Way which is, so to speak, our parish, contains about three hundred thousand million stars. There are many million assemblages similar to the Milky Way, and the distance from one such assemblage to the next takes about two million years for light to traverse.' He goes on to mention that the sun weighs about two billion billion billion tons, and that the Milky Way weighs about a hundred and sixty thousand million times as much as the sun. Yet, though there is a good deal of matter in the universe, 'the immensely larger part of it is empty, or very nearly empty'.

In recent years the revelations of the astronomers have abashed our sense of importance. There was a time when the earth was regarded as the centre of the universe, and man appointed there in the image of the Creator. Now it would seem that no words can express his insignificance in the theatrical enormity of the firmament. He is but a speck of dust on a planet which is as a grain of sand in the countless constellations, and the sun but as a firefly amidst the multitudinous orbs which themselves are lost in the immensity of the void.

But we are coming to see that this attitude is absurd. What is the imperial magnitude of these galaxies in comparison with the human consciousness? They might just as well be toy balloons. St Augustine declared that the human mind was of greater dignity than the entire inanimate creation, and can we say that he is out of date in that assertion? For these systems slung aloft the vast abortive voids have nothing but their bulk to recommend themselves and to astonish us. They are unaware of themselves and of each other. They cannot think. They cannot feel. They cannot see anything. A worm is a far grander thing than the insensate rock on the unseen precipice in unknown regions of the universe. Can a place without spectators properly claim existence? Remove thought, our thought, and what is left? Everything vanishes. It may be there. But since there are no spectators to witness its existence, it might just as well not exist. In his recent book, *The Myth of the Machine*, (another mine of knowledge and insight), Lewis Mumford observes that the immensities of space and time with which science now daunts us turn out as 'quite empty conceits except as related to man'. He emphasizes that for man to feel belittled, as many apparently do, by the vastness of the universe 'is precisely like being frightened by his own shadow. It is only through the light of consciousness that the universe becomes visible, and should that light disappear only nothingness would remain. Except on the lighted stage of human consciousness, the mighty cosmos is but a mindless nonentity. Only through human words and symbols, registering human thought, can the universe disclosed by astronomy be rescued from its everlasting vacuity. Without that lighted stage, without the human drama played upon it, the whole theatre of the heavens, which so deeply moves the human soul, exalting and dismaying it, would dissolve again into its own existential nothingness, like Prospero's dream world.' Immanuel Kant held that time and space are only mental attributes. It is easy to see that we have invented time but not so easy to see how we can have invented space. But Mumford helps me to some understanding of it when he remarks that 'without man's time-keeping activities, the universe is yearless, as without his spacial conceptions, without his discovery of forms, patterns, rhythms, it is an insensate, formless, timeless, meaningless void.'

Mankind is not a miserable adjunct on the surface of an uncaring universe. It encompasses us: but we, by thought, encompass it. The universe is aware of nothing. We alone give it significance. We only are awake. We only are aware. Our position is central and godlike. Would that we could be worthy of it!

It is customary these days to say that probably all the galaxies are not necessarily a dead loss, but could have life upon them; that on many other planets beings in some respects similar to ourselves are existing. That could be so. It may well be so. But it is just as likely that we are the only conscious beings in the entire universe.

The reader may have noticed that I enjoy the company of the philosophers in so far as I can follow them. But this does not include the new materialists of our day, who are more or less impossible to understand. They tell us severely that we really must drop the concept of mind. We can quite well do without it. Better to think in terms of chemistry and vibrations, and not bother about consciousness or the self. True, there are perceptions and events taking place within us, but there is no actual self, only a series of fleeting impressions, fancies, sensations, pains and pleasures which succeed each other rapidly. The whole thing is determined and automatic. The notion of self is an hallucination. We have bodies consisting of tubes, glands, canals, lungs, heart, brains, vascular and other systems, springs, levers, muscles – all getting on like clockwork without benefit of metaphysical explanations.

These writers are sometimes called rationalists: as if irrationality could go further! I prefer the term reductionists, for they reduce everything to a 'merely' this, or 'merely' that; merely a machine, merely automatic, merely chance, merely an accident bound to happen given enough time. As William James put it: 'What is higher is explained by what is lower, and treated forever as a case of "nothing but" – nothing but something else of an inferior sort.'

It is a rather interesting approach. Mind discovers that there is no mind. A ghost detects itself as a phantom. An hallucination stumbles upon an hallucination. There is an object but no subject. The obviously relevant is irrelevant. Our consciousness of consciousness is a delusion. A mechanism announces that it is mechanical.

But it is no good. Mr Jones knows that he is Mr Jones. And he can recognize the existence of Mr Brown. Oxygen, hydrogen, carbon, water, lead, stone, electrons, protons, or any combinations of such cannot become conscious of themselves or of others: carbon cannot say: 'Look, here comes hydrogen.' Because of this Mr Jones, in comparison, is as a god looking down from Olympian heights. For consciousness can neither be explained nor set aside. I quote again from Macneile Dixon whose great book *The Human Situation* never goes out of print. 'How we arrived on Olympus, on this height from which all the kingdoms of heaven and earth may be surveyed, I do not profess to tell you. Ask the space-time philosophers, or the physiologists or bio-chemists. Perhaps the brain secretes this magical essence, consciousness, as the liver secretes bile. Ask those who are prepared to explain the process to you, in this or in some other way. You may chance to find, even among philosophers, people who see nothing remarkable about consciousness. For my part I hold that neither intellect nor imagination, neither science nor logic, can cross the threshold of this mystery, nor language lay hold of it. To regard the advent of consciousness, that is, the world's coming to a knowledge of itself, the awakening of a soul in nature, to take this unexampled overwhelming fact as of course and for granted, as no singular event, or anything out of the way noteworthy or surprising, or again as a thing of accident among other accidents, were for me no easier a thought than the notion of the Himalayas giving way to laughter, or the ocean writing its autobiography. When you begin to suppose such things you make a clown of reason and adorn it with cap and bells.'

Nor is it any more sensible to say that 'given time' this or that miracle will eventually come about. Oceans of time would do nothing without an evolutionary creative will – however incomprehensible to us. In the Seventh Section of his compelling dialogue called *Old Rectory, or The Interview*, Martyn Skinner puts the evolutionary point more succinctly than I can in prose.

> . . . For today let's pause
> At my first groping after the First Cause,
> Which led me to acknowledge (groping still)
> That if what once was called primeval slime
> (In current jargon, pre-biotic soup)
> Evolved in course of eons to a group
> Playing Beethoven, it needed more than time
> And chance, it needed a creative will

> To foster that emergence, and express
> Amoeba as A Minor.

It is strange that people should wish to solve 'the riddle of the world'. Because to succeed is so manifestly out of the question. Take the astronomical statement which I have quoted from Bertrand Russell about the stars and the distance between the stars, and add to it if you wish the current statements about the 'receding galaxies' going away somewhere at so many thousand miles a minute or second. Plainly, what is really being proposed is infinite space – that is, space which has no end. But that is something which we cannot conceive. In mid-Atlantic, alone at the prow of a small steamer, with no responsibility for anything – something I enjoy more than anything on earth – I gaze for hours at the horizon. It is not far away. I force my imagination to see beyond it, the water-plain going on for another fifteen hundred miles. An enthralling image, widening the horizon of one's own conceptions. Then I try to think of it going on indefinitely, for ever. There my imagination stops. I cannot conceive everlasting space: it is strictly inconceivable. But finite space is also inconceivable, for immediately one asks what is behind *it*. (I was very disappointed when Einstein came up with *curved* space, as that did nothing for me, since naturally I was constrained to ask what lay beyond the curve.) It is the same with our familiar friend the first cause: we have to ask *its* cause. So also with time: we cannot conceive eternity; it is then that the effort involved does truly make our minds boggle. And one other thing. I have always attached importance to the fact I cannot think of, I cannot conceive – nothing. Nothing is not the same thing as a void; it is nothing at all; and this we cannot imagine. And I have often wondered that if I cannot conceive nothing, then perhaps there is no such thing as nothing, no such thing as a beginning, and no such thing as an end. These are encouraging thoughts really, positive in their negativity. Perhaps at death we might get the surprise of our lives! The poet Shelley couldn't wait to clear the matter up, and several times when he was out in a boat his friends had a job to prevent him from jumping overboard to drown and find the answer.

Personally, I like the incomprehensible. 'The formation of celestial bodies,' said Kant, 'the causes of their movements, and in short, the origin of the entire Cosmos, will be explained sooner than the mechanism of a plant or a caterpillar.' And when scientists tell us that a single germ of life is also a whirlwind of millions of electrons revolving in their orbits millions of times in the millionth of a second, we know that they are referring to the incomprehensible. I cling to this ignorance as the most solid item in all our store of knowledge.

If this sounds too negative I would commend a remark dropped by George Steiner the other day when writing on Heinrich von Kleist. 'Kleist emerged from the *Kant-Krise* convinced that man's fundamental insight into nature and meaning of existence derives not from analytical and objective meditation but from enigmatic lightning-flashes of total experience.' Do we really need more than this? Have we the right to expect more? 'Why should you expect,' asked a distinguished French scientist recently, 'an ex-simian, who only lately climbed down from the trees of his native forest, to understand the Universe?'

I have advanced in these pages my philosophy of biology in terms of creative evolution; the progress from Amoeba to A Minor. There is comfort in an intellectual and scientific scheme that seems to square with the facts and make sense. But, it may fairly be asked, if this is so, if a man is such a unique arrival, if through his consciousness *the world comes to a knowledge of itself*, how to account for the present state of affairs? How is it that he does not seek, day and night, to increase this consciousness, to acknowledge this uniqueness, and to curb those primitive instincts which his cleverness has made so disastrous? How comes it that he prefers to visit the moon, to spoil nature, and to use his incredible expertise to carry on fighting himself for possession of the earth? I have spoken of the miraculous birth of language. But are we not now busy debasing it to such an extent that in the English language alone a phrase such as 'I'll take care of him' means 'I'll kill him', while endless webs of words are used not for communication but for the concealment of truth, or sheer babbling, so that there is real pertinence in the bitter lines by Colin Hurry in his *Receding Galaxy*:

> In the beginning chaos, then the Word.
> Then light; and life upon the planet stirred.
> Now, aeons after, chaos rules again.
> The Word forgotten, only words remain.

Perhaps the rapid growth of the cortex superimposed upon the older portions of the brain, instead of co-ordinating and 'spiritualizing' our personality, has become an evolutionary disaster.* Yet we know nothing of the primordial energy which has caused evolution. A resurgence of spirit is not less likely than a complete collapse of spirit. History is full of surprises. Who could have foretold the Crusades or the Renaissance? There may come a new form of Crusade in reaction against present action. Always going to extremes, mankind may take a turn for the good in altogether unforeseen excess of energy.

All the same, we can face life, if we choose, without being parasitic either upon a scientific theory, a theological dogma, or sociological hopes and fears. There is the poetic approach to reality. I favour it myself. There are many good reasons why thousands, or rather millions, of people cannot possibly accept it as meaning anything to them. On the other hand there are thousands (not millions) who are satisfied with the poetic approach to reality without any idea that it qualifies for so grand a definition. For it is scarcely more than an attitude of mind. I would define it as an acceptance of life based upon the

*See page 79.

187

experience of beauty and the intuition of the *intrinsic value* of the incomprehensible.

'The trouble is,' said John Galsworthy rather surprisingly, 'The trouble is that there are not enough lovers of beauty in the world.' A profound remark surely, and pleasantly short-winded as a philosophy of life, for none of us really can think in terms of five hundred pages. I am particularly drawn to a remark made by Chekhov in a book so obscure that not one in a million readers has a chance of seeing a copy. 'So long as a man likes the splashing of a fish,' he wrote, 'he is a poet. But when he knows that the splashing is nothing but the chase of the weak by the strong, he is a thinker; but when he does not understand what sense there is in the chase, or what use in the equilibrium which results from destruction, he is becoming silly and dull as he was when a child. And the more he knows and thinks the sillier he becomes.' Chekhov was there being perhaps a bit patronizing with regard to the poet, for the poet is also the thinker. In fact the poet represents Goethe's man in another observation I am fond of quoting. 'Awe is the highest thing in man,' he said. 'And if the pure phenomena awaken awe in him he should be content; he can be aware of nothing higher and should seek nothing beyond: here is the limit.' That very well defines what I mean by the poetic approach to reality. The reason why I admit that it can have only a limited appeal is well summed up in the few words with which Goethe concluded his apopthegm: 'But for most men the vision of the pure phenomena is not enough, they insist upon going further like children who peep in a mirror and then turn it round to see what is on the other side.' This is inevitable. The desire to explain the intuitions arising from emotional experience, and to give them a scientific or a theological gloss, is so deep-rooted that it is almost a necessity for the majority of us. And terms such as humanist, agnostic, atheist, are very cold (though the last-named only means non-theist). If the word had not been used in another context, I would prefer to call myself an existentialist – one who is grateful just for having been given the opportunity of existing.

There is this to be said: if we allow our sense of beauty to grow, we may come to find that the peculiar assurance that accompanies it, in spite of the dark side of life, is so great as to *obliterate* all other considerations. The finished article is so extraordinary that the means by which it came into existence is of secondary concern. That is what I had in mind when I said earlier: Well done atoms! Well done mechanism! Well done accident!

Actually, very few people are blind to beauty. Let a beautiful woman walk through a crowded room and all eyes will turn and gaze at her. The experience of beauty is also the experience of love. Love is not blind: it is the power to see the beauty in another person. And the case for cultivating the experience of beauty is stronger than is generally recognized. It is strong for this reason: that if we are not concerned about beauty, and mind about it, we will not be concerned about ugliness either, and will not mind about it; when we cease to be aware of beauty we cease to be aware of ugliness; when we no longer worship beauty we seem positively to worship ugliness! And the awful thing is that while beauty can inspire us, can lift us high in apprehension, ugliness can *frighten* us and cast us down.

> For ugliness, when it is harsh, extreme,
> Has power to terrify, and through the sense
> Work like a visionary experience.*

I have tried to cultivate my sense of 'awe before the pure phenomena'; but I hadn't bothered to look at the body! When St Augustine in his *Confessions* thought about this same lapse he declared 'much wonder groweth upon me, yea astonishment seizeth me'. Men go abroad, he said 'to wonder at the height of mountains, at the huge waves of the sea, at the long courses of the rivers, at the vast compass of the ocean, at the

Letters to Malaya, Book III, Martyn Skinner.

189

circular motion of the stars; and they *pass by themselves without wondering.*' I have tried to make good, for myself at any rate, this strange omission, and have been continuously amazed by the intricacy and perfection of our interior apparatus, the operations of the lungs and the heart, the consummate exactness of intercommunication between the vascular and lymphatic systems, not to mention the endless muscular devices by which the body maintains its existence. It seems a marvel that such an elaborate affair could keep going for more than an hour or two without some mishap. The lungs! Fancy being able to breathe every minute year after year without the lungs ever getting stuck. I began this Discourse thinking about *breathing*, and I end again thinking of it, with amazement. We wonder, or we ought to wonder, how few car accidents there are, seeing that a cataract of vehicles flows through the roads every hour of every day; but you can drive from Cornwall to Scotland without seeing a single collision. It is more extraordinary still to be able to go from one end of London to the other, along crowded pavements, without seeing anyone collapse. You wouldn't credit it, until demonstrated; and then we do not credit it, but take it for granted! Thinking on these things, after my studies, I appreciate and echo the words of Novalis: 'There is but one temple in the universe, and that is the Body of Man. Nothing is holier than that high form. Bending before man is a reverence done to this Revelation in the Flesh. We touch heaven when we lay our hands on a human body.'

Bibliography

ADRIAN, E. D. *The Basis of Sensation*. London, 1928

ALLEN, E. *Sex and Internal Secretions*. Baltimore, 1939

ALVAREZ, W. C. *The Mechanics of the Digestive Tract*. New York, 1940

ANDREWS, M. *The Life that Lives on Man*. London, 1976

BARCROFT, H. and SWAN, H. J. C. *Sympathetic Control of Human Blood Vessels*. London, 1953

BEACH, F. A. *Hormones and Behaviour*. New York, 1949

BERKELEY, George. *Treatise on the Principles of Human Knowledge*. 1710.

BLAKEMORE, C. *Mechanics of the Mind*. Reith Lectures 1976

BRADY, J. V. *Biological and Biochemical Basis of Behaviour*. Wisconsin, 1958

BRAIN, R. *The Nature of Experience*. Oxford, 1959

BRIGGS, R. and MACFARLANE, R. G. *Human Blood Coagulation and its Disorders*. London, 1953

BROWN, Christy. *My Left Foot*, Foreword and Epilogue by Dr Robert Collis

BURROWS, H. *Biological Action of Sex Hormones*. Oxford, 1949

BUTLER, Samuel. *Evolution Old and New. God the Known and the Unknown*. London

CAMPBELL, E. J. M. *The Respiratory Muscles and Mechanics of Breathing*. Chicago, 1958

CANNON, W. B. *Bodily Changes in Pain, Hunger, Fear and Rage*. New York, 1929

CROW, W. R. *A Synopsis of Biology*. Bristol, 1960

DIXON, W. MACNEILE. *The Human Situation*. Oxford

DURANT, William. *The Story of Philosophy*. London, 1926

ECCLES, J. C. *The Physiology of Nerve Cells*. Baltimore, 1957

FABRE, Jean-Henri. *Marvels of the Insect World*. London

FROMAN, R. *The Many Human Senses*. London, 1969

FULTON, J. F. *Muscular Contraction of the Reflex Control of*

Movement. Baltimore, 1926

FULTON, J. F. *Functional Localization in the Frontal Lobes and Cerebellum.* Oxford, 1949

FULTON, J. F. *Physiology of the Nervous System.* Oxford, 1949

GELDARD, F. A. *The Human Senses.* New York, 1953

GROSS CLARK, W. E. LE *The Tissues of the Body.* Oxford, 1958

HARRIS, G. W. *Neural Control of the Pituitary Gland.* London and Baltimore, 1955

HARRISON, Michael. *Fire from Heaven.* London, 1976

HARTMAN, C. G. *Time of Ovulation in Women: a Study on the Fertile Period in the Menstrual Cycle.* Baltimore, 1936

HAWKES, J. *Man on Earth.* London, 1954

HEAD, H. *Studies in Neurology.* Oxford, 1920

HENDERSON, L. J. *Blood: a Study in General Physiology.* Yale, 1928

HILL, A. *The Body at Work.* London, 1910

HOOKE, Robert. *Microphalia.* 1660

HUME, David. *Treatise on Human Nature.* 1739

HUXLEY, T. H. *Lessons in Elementary Physiology.* London, 1893

HUXLEY, J. *The Uniqueness of Man.*

HUDSON, W. H. *The Book of a Naturalist.* London, 1936

HORROBIN, D. F. *An Introduction to Human Physiology.* Lancaster, 1973

KANT, Immanuel. *The Critique of Pure Reason. The Critique of Judgement.*

KEITH, A. *Elements of Physiology (Home University Library).* 1926

KELLER, H. *The World I Live In.* London, 1908

KINGLAKE, A. W. *Eothen.* London

LASLETT, P. Editor of BBC Lectures on Perception under title of The Physical Basis of Mind. Oxford, 1950

LEWIS, T. *The Blood Vessels of the Human Skin and their Responses.* London, 1927

LEWIS, T. *Pain.* New York, 1942

LOCKE, John. *Essay Concerning Human Understanding.* 1690

LUCKIESH, M. and MOSS, F. K. *The Science of Seeing.* New York, 1937

LUSK, G. *The Elements of the Science of Nutrition.* Philadelphia, 1928

McDOWALL, R. J. S. *The Control of the Circulation of the Blood.* London, 1938

MACHMAN, SOHN D. *Chemical and Molecular Basis of Nerve Activity.* New York, 1959

MANN, I. and PIRIE, A. *The Science of Seeing.* London, 1946

MUMFORD, L. *The Myth of the Machine.* London, 1967

NOVALIS. the pen-name of Frederich von Hardenberg. 1772 (See Carlyle's *Miscellaneous Essays,* Vol. II)

PAINE, Thomas. *The Age of Reason.* London, 1881

PICKERING, J. W. *The Blood Plasma in Health and Disease.* London, 1928

PAVLOV, I. P. *The Work of the Digestive Glands.* London, 1910

RAJAH OF AUNDH. *The Ten-point Way to Health.* London

RASMUSSEN, A. T. *The Principal Nervous Pathways.* New York, 1941

REYNOLDS, S. R. M. *Physiology of the Uterus.* New York, 1948

RÉVÉSZ, G. *The Origins and Prehistory of Language.* New York and London, 1956

RUCH, T. C. and FULTON, J. F. *Medical Physiology and Biophysics.* Philadelphia and London, 1960

RUSKIN, John. *The Eagle's Nest.*

SAPIR, E. *An Introduction to the Study of Speech.* New York and London, 1921

SHAW, BERNARD. *Preface to Saint Joan.* London *Preface to The Miraculous Birth of Language.*

SHERRINGTON, C. *The Integrative Action of the Nervous System.* Yale, 1947

SHOLL, D. A. *The Organisation of the Cerebral Cortex.* London, 1956

SKINNER, Martyn. *Letters to Malaya.* London

SMITH, A. *The Body.* London, 1968

SMITH, H. W. *The Kidney, its Structure and Function in Health and Disease.* New York and Oxford, 1951

SUTTON-VANE, S. *The Story of Eyes.* 1958

TYNDALL, John. Address to the British Association, Belfast. London, 1874

VERZAR, F. and MCDOUGALL, E. J. *Absorption from the Intestine.* New York and London, 1936

VAN DER POST, L. *Jung and the Story of our time.* Hogarth Press, 1976

WALKER, A. E. *The Primate Thalmus.* Chicago, 1938

WALTER, Grey. *The Living Brain,* London

WHORF, B. L. *Language, Thought, and Reality.* Massachusetts, 1956

WILSON, R. A. *The Miraculous Birth of Language.* London, 1942
All readers acquainted with Professor Wilson's celebrated book (with a Preface by Bernard Shaw in 1942) will observe my indebtedness to him.

Students who wish to widen the angle or make qualifications can consult Révész, Sapir, and Whorf.

WOLF, S. and WOLF, H. G. *Human Gastric Function.* New York and Oxford, 1947

YOUNG, J. Z. *Doubt and Certainty in Science.* Oxford, 1960

The author is grateful to the Cresset Press for permission to quote 'The Primordial Cell' from *The Spirit Watches* by Ruth Pitter.

Index